CURRENT

ARRIVAL OF THE FITTEST

ARRIVAL

of the

FITTEST

SOLVING EVOLUTION'S
GREATEST PUZZLE

Andreas Wagner

CURRENT

CURRENT
Published by the Penguin Group
Penguin Group (USA) LLC
375 Hudson Street
New York, New York 10014

USA | Canada | UK | Ireland | Australia | New Zealand | India | South Africa | China
penguin.com
A Penguin Random House Company

First published by Current, a member of Penguin Group (USA) LLC, 2014

Illustrations by the author

LIBRARY OF CONGRESS CATALOGING-IN-PUBLICATION DATA
Wagner, Andreas, 1967 January 26–
Arrival of the fittest : solving evolution's greatest puzzle / Andreas Wagner.
pages cm
Includes bibliographical references and index.
ISBN 978-1-59184-646-8
1. Natural selection. 2. Evolutionary genetics. I. Title.
QH375.W327 2014
572.8'38—dc23   2014009774

Printed in the United States of America
10 9 8 7 6 5 4 3 2 1

Set in Minion Pro with Amasis MT Std
Designed by Daniel Lagin

CONTENTS

CONTENTS

ARRIVAL OF THE FITTEST

PROLOGUE

World Enough, and Time

In the spring of 1904, Ernest Rutherford, a thirty-two-year-old New Zealand–born physicist then working at McGill University in Canada, gave a lecture at the world's oldest scientific organization, the Royal Society of London for Improving Natural Knowledge. His subject was radioactivity and the age of the earth.

At that time, scientists had long since forsworn the biblical accounts asserting that the earth was only six thousand years old. The most widely accepted dates had been calculated by another physicist—William Thomson, better known as Lord Kelvin—who had used the equations of thermodynamics and the earth's heat conductivity to estimate that the planet was somewhere around twenty million years old.

In geology, that's not a lot of time, and the implications were profound. The earth's geological features could not have appeared within this duration if processes like volcanism and erosion proceeded at today's rate.[1] But the real victim of Kelvin's estimate was Charles Darwin's theory of evolution by natural selection. Darwin had described himself as "greatly troubled at the short duration of the world according to Sir W. Thomson."[2] He knew that organisms had not changed much since the last ice ages, and from

1

such little change he inferred that the amount of time needed to create all organisms—alive today or preserved in fossils—must be truly enormous.[3] Twenty million years was not enough time to create life's diversity.

But Rutherford, who had discovered the phenomenon of radioactive half-life only a few years before, knew that Kelvin was wrong, by at least several orders of magnitude. As he later recalled:

> I came into the room, which was half dark, and presently spotted Lord Kelvin in the audience and realized that I was in for trouble at the last part of the speech dealing with the age of the earth, where my views conflicted with his. . . . The discovery of the radio-active elements, which in their disintegration liberate enormous amounts of energy, thus increases the possible limit of the duration of life on this planet, and *allows the time claimed by the geologist and biologist for the process of evolution.*[4] (emphasis added).

And that was that. Kelvin died in 1907. Rutherford won the Nobel Prize in 1908, and by the 1930s his radiometric methods had shown that the earth was around 4.5 billion years old. Darwin's theory was saved, since the processes of random mutation and selection now had the time needed to create life's enormous complexity and diversity.

Or did they?

Consider the peregrine falcon, *Falco peregrinus,* one of nature's great predators and an organism of marvelous perfection. Its powerful musculature, matched with an extremely lightweight skeleton, makes it by far the fastest animal on earth, able to reach more than 200 miles per hour in one of its characteristic dives. All that speed translates into enormous kinetic energy when the falcon strikes its prey in midair with razor-sharp talons. If that impact alone does not deliver death, the falcon can sever the spinal column of its prey with a conveniently notched upper beak.[5]

Before moving in for the kill, *F. peregrinus* needs to track down its prey. The targeting mechanism is a pair of eyes with full-color binocular vision, possessing resolving power more than five times greater than a human's, which means that a peregrine can see a pigeon at distances of more than a mile.[6] Like many predators, the falcon has an eye with a nictitating membrane—a third eyelid—a bit like a windshield wiper that removes dirt while keeping the eye moist during a high-speed chase. The falcon's eyes also harbor more photoreceptors, the rods that capture images in very low light, and the cones that provide color vision.[7] Its photoreceptors render even long-wavelength ultraviolet light visible.

A marvel indeed. But even more marvelous is knowing that every one of those brilliant adaptations is the sum of innumerable tiny steps, each one preserved by natural selection, each one a change in a single molecule. The deadly beak and talons of *F. peregrinus* are built from the same raw material as its feathers, the protein molecules known as keratin, the human versions of which make up your hair and nails.[8] For color vision, those extraordinary eyes depend on opsins, protein molecules in the eyes' rods and cones. Crucial for their remarkable acuity are their lenses, composed of transparent proteins known as crystallins.[9]

The first vertebrates to use crystallins in lenses did so more than five hundred million years ago, and the opsins that enable the falcon's vision are some seven hundred million years old.[10] They originated some three billion years after life first appeared on earth. That sounds like a helpfully long amount of time to come up with these molecular innovations. But each one of those opsin and crystallin proteins is a chain of hundreds of amino acids, highly specific sequences of molecules written in an alphabet of twenty amino acid letters. If only one such sequence could sense light or help form a transparent cameralike lens, how many different hundred-amino-acid-long protein strings would we have to sift through? The first amino acid of such a string could be any one of the twenty kinds of amino acids, and the same holds for the second amino acid. Because $20 \times 20 = 400$, there are there are 400 possible strings of two amino acids. Consider also the third amino acid, and you have arrived at $20 \times 20 \times 20$, or 8,000, possibilities. At four amino acids we already have 160,000 possibilities. For a protein with a hundred amino acids (crystallins and opsins are much longer), the numbers multiply to a 1 with more than 130 trailing zeroes, or more than 10^{130} possible amino acid strings. To get a sense of this number's magnitude, consider that most atoms in the universe are hydrogen atoms, and physicists have estimated the number of these atoms as 10^{90}, or 1,000,000,000,000,000,000,000,000,000,000,000, 000,000,000,000,000,000,000,000,000,000,000,000,000,000,000,000, 000,000. This is "only" a 1 with 90 zeroes. The number of potential proteins is not merely astronomical, it is hyperastronomical, much greater than the number of hydrogen atoms in the universe.[11] To find a specific sequence like that is not just less likely than winning the jackpot in the lottery, it is less likely than winning a jackpot every year since the Big Bang.[12] In fact, it's countless billions of times less likely. If a trillion different organisms had tried an amino acid string every second since life began, they might have tried a tiny fraction of the 10^{130} potential ones. They would never have found the one opsin string. There are a lot of different ways to arrange molecules. And not nearly enough time.

When the seventeenth-century lyric poet Andrew Marvell bemoaned, "Had we but world enough, and time" to avoid the "deserts of vast eternity" that lay before him, he was attempting to unlock his mistress's bedchamber, not the secrets of nature. But he was on to something. Common wisdom holds that natural selection, combined with the magic wand of random change, will produce the falcon's eye in good time. This is the mainstream perspective on Darwinian evolution: A tiny fraction of small and random heritable changes confers a reproductive advantage to the organisms that win this genetic lottery and, accumulating over time, such changes explain the falcon's eye—and, by extension, everything from the falcon itself to all of life's diversity.

The power of natural selection is beyond dispute, but this power has limits. Natural selection can *preserve* innovations, but it cannot *create* them. And calling the change that creates them random is just another way of admitting our ignorance about it. Nature's many innovations—some uncannily perfect—call for natural principles that accelerate life's ability to innovate, its *innovability*.

For the last fifteen years, I have been privileged to help uncover these principles, first in the United States and later, joined by a group of highly talented researchers, in my laboratory at the University of Zürich in Switzerland. Using experimental and computational technologies unimagined by Darwin or Rutherford, our goal is not to discover individual innovations, but to find the wellsprings of *all* biological innovation. What we have found so far already tells us that there is much more to evolution than meets the eye. It tells us that the principles of innovability are concealed, even beyond the molecular architecture of DNA, in a hidden architecture of life with an otherworldly beauty.

These principles are the subject of this book.

CHAPTER ONE

What Darwin Didn't Know

Sallie Gardner was the world's first movie star. Her graceful debut in 1878 launched cinema itself, though she was only six years old. Sallie, you see, happened to be the Thoroughbred horse that the English-born photographer Eadweard Muybridge shot in full gallop with his zoopraxiscope, an array of twenty-four cameras along her path, to settle a pressing question that undoubtedly keeps many people awake at night: Does a galloping horse ever lift all four legs off the ground? (The answer is yes.) His grainy, jerky silent movie, all of a second long, is worlds apart from the high-definition digital surround-sound cinematography taken for granted in the early twenty-first century. Yet the time separating Muybridge's photographic study from modern movies spans just over a century, a stretch not much longer than the time since Darwin published *The Origin of Species,* only nineteen years before Sallie Gardner's star turn.

During the same time, biology has been transformed by a revolution even more dramatic than the cinematic one.[1] This revolution has revealed a world as inaccessible to Darwin as outer space was to cavemen. And it has helped to answer the single most important question about evolution, the question that Darwin and generations of scientists after him did not, could not touch: How does nature bring forth the new, the better, the superior? How does life create?

You might be puzzled. Wasn't that exactly Darwin's great achievement, to understand that life evolved and to explain how? Isn't that his legacy? Yes and no. Darwin's theory surely is the most important intellectual achievement of his time, perhaps of all time. But the biggest mystery about evolution eluded his theory. And he couldn't even get close to solving it. To see why, we first need to take a look at what Darwin knew and what he didn't, what was new about his theory and what wasn't, and why only now, more than a century later, can we begin to see how the living world creates.

Germs of thought about an evolving natural world existed long before Darwin. No fewer than twenty-five hundred years ago, the Greek philosopher Anaximander—better known as the great-grandfather of the heliocentric worldview—thought that humans emerged from fish. The fourteenth-century Muslim historian Ibn Khaldun thought that life progressed gradually from minerals to plants to animals. Much later, the nineteenth-century French anatomist Etienne Geoffroy Saint-Hilaire deduced from fossilized reptiles that they had changed over time.[2] The Viennese botanist Franz Unger argued in 1850, just a few years before Darwin published *The Origin of Species* in 1859, that all other plants descend from algae.[3] And the French zoologist Jean-Baptiste Lamarck postulated that evolution occurred through use and disuse of organs. Some of the earliest thinkers even seem prescient about evolution, until you dig a bit and find some bizarre nuggets, such as Anaximander's notion that early humans lived *inside* fish until puberty, when their hosts burst and released them. Beliefs that are alien to today's science persisted well into Darwin's era. According to one of them, shared by many from the ancient Greeks to Lamarck, simple organisms are spontaneously created from inanimate matter like wet mud.[4]

Just as evolution had its proponents, it had equally vocal opponents well into Darwin's era. And no, I do not mean people like today's young earth creationists—half literate and wholly ignorant—who believe that earth was created on a Saturday night in October of 4004 BC (and that Noah's Ark could have saved more than a million species, but Noah somehow forgot the huge dinosaurs, perhaps forgivably so, considering that he was six hundred years old). I mean scientific leaders of the time. One of them was the French geologist Georges Cuvier, the founder of paleontology, literally the science of "ancient beings" (think dinosaurs).[5] He discovered that the fossils embedded in older rocks are quite different from those in younger rocks, which resemble today's life. Yet he thought that each species had essential, immutable characteristics, and could only vary in superficial traits. Another example is Carl Linnaeus, who lived a mere century before Darwin. He is the father of our modern system for classifying life's diversity, yet until late in life he did not believe in evolution's great chain of living beings.[6]

Christian beliefs are the best-known reason for such resistance. To Cuvier, life's diversity wasn't evidence of evolution but of the Creator's great talents. Another reason, however, has even deeper roots. It goes back all the way to the Greek philosopher Plato, whose influence on Western philosophy is so great that the twentieth-century philosopher Alfred North Whitehead demoted all of European philosophy to "a series of footnotes to Plato."[7] Plato's philosophy was deeply influenced by the ideal, abstract world of mathematics and geometry. It maintains that the visible, material world is but a faint, fleeting shadow of a higher reality, which consists of abstract geometric forms, such as triangles and circles. To a Platonist, basketballs, tennis balls, and Ping-Pong balls share an *essence,* their ball-like shape. It is this essence—perfect, geometric, abstract—that is real, not the physical balls, which are as fleeting and changeable as shadows.

The goals of scientists like Linnaeus and Cuvier—to organize the chaos of life's diversity—are much easier to achieve if each species has a Platonic essence that distinguishes it from all others, in the same way that the absence of legs and eyelids is essential to snakes and distinguishes them from other reptiles. In this Platonic worldview, the task of naturalists is to find the essence for each species. Actually, that understates the case: In an essentialist world, the essence really *is* the species.[8] Contrast this with an ever-changing evolving world, where species incessantly spew forth new species that can blend with each other.[9] The snake *Eupodophis* from the late Cretaceous period, which had rudimentary hind legs, and the glass lizard, which is alive today and lacks legs, are just two of many witnesses to the blurry boundaries of species. Evolution's messy world is anathema to the clear, pristine order that essentialism craves. It is thus no accident that Plato and his essentialism became the "great antihero of evolutionism," as the twentieth-century zoologist Ernst Mayr called it.[10]

In the controversy between Darwinists and their opponents, fossils like *Eupodophis* were mere boulders in a mountain of evidence that helped Darwin's supporters gain the upper hand.[11] At Darwin's time, systematists had already classified thousands of living species and unveiled deep similarities among them. Geologists had discovered that the earth's surface was roiling, incessantly creating, folding, and crushing layers of rock. Paleontologists had discovered countless extinct species, some in young rocks and similar to the life we know, others in ancient rocks and very different. Embryologists had shown that organisms as different as a freely paddling shrimp and a barnacle clamped to a ship's hull can have deeply similar embryos.[12] Explorers, Darwin among them, had found many intriguing patterns of biogeography. Small islands have fewer species, opposite shores of the same continent harbor very different faunas, Europe and South America host completely different mammals.[13]

A special creation of each species would leave all these threads of knowledge in a messy tangle. Darwin, one of the greatest synthesizers of all time, wove them into the beautiful fabric of his theory. He threw the gauntlet at creationists by claiming that *all* life shares a common ancestor, and thereby dismissed biblical Genesis from the debate table.

That was Darwin's first great insight. The second one was the central role of natural selection, an insight inspired by the spectacular success of animal and plant breeders.[14] The *Origin*'s entire first chapter marvels at the diversity of domestic dogs, pigeons, crop plants, and ornamental flowers that human breeders had produced. It is indeed stunning to think that humans could create Great Danes, German shepherds, greyhounds, bulldogs, and Chihuahuas, all from a common lupine ancestor, and all within mere centuries. Darwin realized that natural selection is not so different from such human selection, except that it operates on a much grander scale, and over eons of time. Nature incessantly creates new variants of organisms, most inferior, a few of them superior, and all of these variants must pass through the sieve of natural selection. Only individuals best adapted to their environment survive, procreate, and give rise to further variants. Given enough time, this process helps explain all of life's diversity, so much so that the geneticist Theodosius Dobzhansky could say in 1973 that "nothing in biology makes sense except in the light of evolution."

From the very beginning, that light shone more brightly on some of life's mysteries than on others. One of them was left in especially deep shadows: the mechanism of heritability. Without some mechanism that guarantees faithful inheritance from parents to offspring, adaptations— a bird's wing, a giraffe's neck, a snake's fangs—cannot persist over time. And without inheritance, selection would be powerless. Darwin himself had no idea why children resemble their parents, and his frankness in admitting ignorance is disarming. "The laws governing inheritance are for the most part unknown," he said in the *Origin*.[15]

Darwin's theory was a bit like that first movie of a galloping horse, revolutionary when compared to still photography, but only a modest step on the path to full-length feature films. The next step on biology's path—explaining inheritance—was already made by the time Darwin died, but he did not know it. Nor did any other prominent scientist, although decisive experiments had already started in 1856, three years before Darwin published the *Origin*.[16] Even the scientist who performed these experiments would not live to see the avalanche of progress he triggered, which would eventually engulf all of biology.

That scientist was the Austrian monk Gregor Mendel, who studied in Vienna and entered St. Thomas Abbey in Brno, where he would experiment on more than twenty thousand pea plants before he became abbot. For his experiments, he deliberately chose pea plants that differed in several discrete features: One plant might produce smoothly round yellow peas, whereas the other would produce wrinkly green peas, but none with in-between color or shape. Other pea plants differed sharply in flower color, pod shape, or stem length. Mendel cross-fertilized these plants and analyzed their offspring, thousands and thousands of plants.

What he saw was that these features often do not blend in the offspring.[17] The offspring's first or second generation produces either round or wrinkly peas, but none with an intermediate shape. And different features can be inherited independently, such that the offspring might sport combinations—round and green, wrinkly and yellow—that neither parent harbored. The causes of inheritance behaved like discrete and indivisible particles. Each parent carried two particles responsible for traits like roundness or color, but would pass only one of them on to its offspring. Different features were inherited through different kinds of particles, and could thus combine and recombine independently.

Mendel worked in an academic backwater far from the intellectual currents of his time. And he committed the error that snuffed many an academic career, then and now: He published little and in the wrong place—in his case a local naturalist journal.[18] And as bad luck would have it, the abbot who succeeded him would burn Mendel's papers after his death. But thirty-four years after its publication in 1865, Mendel's sleeping beauty of a discovery would be roused by the Dutch botanist Hugo de Vries, who independently performed experiments similar to those of Mendel. Historians still argue whether he truly rediscovered Mendel's laws, or whether he learned about Mendel's work during his own experiments and tried to hide his knowledge.[19] The searing disappointment of having not just been scooped, but scooped by three decades, could certainly explain the impulse to rewrite history. Be that as it may, rediscovered Mendel's laws were, and from then on they spread like wildfire. They became the basis of a whole new branch of biology, the science of genetics. Traits that behave as Mendel described exist in many plants and animals, including humans. Some of our Mendelian traits are as odd as the consistency of ear wax (wet or dry), but others are as important as the major blood groups (A or B), or diseases like sickle-cell anemia.

As it turns out, de Vries was to receive a consolation prize. He is the grandfather of the word *gene,* whose importance endures in both science and popular culture. De Vries had called the particles of inheritance that Mendel had described "pangenes," and a few years later the Danish geneticist Wilhelm Ludvig Johannsen would simply drop the "pan."[20]

Johannsen contributed two further important words to the language of modern biology. He coined the word *genotype* and distinguished it from the *phenotype.* In today's language, a genotype comprises all genes of an organism, all its DNA, whereas the phenotype comprises everything else you could observe about the organism: its size, its color, whether it has a tail, or feathers, or a carapace. To see this distinction is crucial, because it allows us to tell cause from effect when organisms change. Take the word *mutation,* which was already used two hundred years earlier for any dramatic change in an organism's appearance. In the early twentieth century, it was sometimes applied to Mendel's units of inheritance, and sometimes to the organism (phenotype), leading to endless confusion about causes and effects of change.[21] A century later we know that mutations change a genotype, like the mutations that altered the blueprint of light-sensing opsin proteins in some of our distant animal ancestors. Such genotypic change can cause changes in a phenotype, and some changed phenotypes become innovations—novel and useful features—like our ability to see the world in color.

Only once we have distinguished between genotype and phenotype can we ask a question crucial to understand life's innovability: *How* do mutations cause changes in phenotypes and bring forth innovations? Because that was the other great mystery left unanswered at the time of Darwin's death: Where do innovations come from? Where do the new variants come from that selection needs? And especially those variants that improve an organism, help it survive a little longer, appear sexier to a mate, or have more babies? One could answer this question with a vacuous platitude: New variants arise randomly, by chance. This platitude is still used today, but Darwin was already familiar with it. And he knew that it explains exactly nothing. He opened the chapter on laws of variation in the *Origin* like this:

> I have hitherto sometimes spoken as if the variations . . . had been due to chance. This, of course, is a wholly incorrect expression, but it serves to acknowledge plainly our ignorance of the cause of each particular variation.

This is not a small problem, because natural selection is not a creative force. It does not innovate, but merely selects what is already there. Darwin realized that natural selection allows innovations to spread, but he did not know where they came from in the first place.

To appreciate the magnitude of this problem, consider that every one of the differences between humans and the first life forms on earth was once an innovation: an adaptive solution to some unique challenge faced by a living being. It might have been the challenge of converting the light energy from the sun into living matter. Or the challenge of converting another living thing into food. Or simply of moving from one place to another. Every square meter of the earth's surface, every cubic meter of the oceans, every meadow, forest, and desert, every city and suburb is packed to the limits with organisms, and each organism exhibits countless such innovations. Fundamental ones like photosynthesis and respiration. Protective ones like reptilian scales and insulating feathers. Supportive ones like connective tissue and skeletons. Some are complex, with hundreds of moving parts, others are not. But no matter how large or small, from the ten feet of a blue whale's tail fluke to the ten microns of a bacterium's flagellum, every single one exists because, at some point since life's origin, the right variation occurred.

Selection did not—cannot—create all this variation. A few decades after Darwin, Hugo de Vries expressed it best when he said that "natural selection may explain the *survival* of the fittest, but it cannot explain the *arrival* of the fittest" (emphasis added).[22] And if we do not know what explains its arrival, then we do not understand the very origins of life's diversity.

Life can innovate, it has innovability. What is more, it can innovate while preserving what works through faithful inheritance. It can explore the new while preserving the old. It can be progressive and conservative at the same time. And through the early twentieth century, biologists had no idea how that is possible. As we shall see, there is no way they could have known. Another century of discoveries was needed before the experimental and computational toolbox of biology became powerful enough to tackle this question.

In fact, looking back, it is remarkable that early-twentieth-century scientists could even distinguish genotypes from phenotypes. They were as ignorant about the material basis of Mendelian inheritance as Muybridge was of color photography. It was not even clear whether genes were intangible concepts, like gravity, or physical objects that could be isolated from a body and studied.[23] Only later would it become clear that genes were very physical, lying on chromosomes and consisting of DNA.

Even before the discovery of genes' physical reality, Mendel's discovery fanned the flames under an old controversy that had simmered since Darwin. Discrete, granular, particulate inheritance flies in the face of an obvious fact that all of us are familiar with. If a six-foot-tall man and a five-foot-tall woman have children, then discrete inheritance demands that their children should be as tall as either parent—five or six feet—but never in between.[24] But we all know that the children's heights lie on a continuum, as do the shapes of their faces, the color of their skin, the contours of their bones, and so on. Naturalists since Darwin found such continuous, blending inheritance everywhere around them, in the yield of crops, the weight of eggs, the sizes of leaves—in brief, in most features of organisms.[25] This kind of variation is clearly important in nature.

The controversy raged around the question of which kind of variation, continuous or discrete, was more important for evolution. The naturalist or gradualist school of thought—Darwin was an early adherent—emphasized the small, continuous variation that we see all around us. The other school—"Mendelists," "mutationists," or "saltationists"—believed in the large, discrete variants that Mendel had studied. In a cartoon version of this dispute, a gradualist would imagine that the many petals of a garden rose emerged from its five-petaled ancestors through gradual additions of petals over many generations. A mutationist, on the other hand, would argue that the multifoliate rose could have appeared in a single saltational "macromutation" from this ancestor.[26]

Looking back, this debate seems just as important as the question that kept medieval scholastics busy: How many angels can dance on the head of a pin? But it just about pierced the heart of Darwinism. For the Mendelists believed less in natural selection than in the power of mutations to bring forth new traits. In their view, the real drivers behind life's evolution were large mutations that created individuals far outside the norm of their species. "Hopeful monsters" is what the German-born zoologist Richard Goldschmidt would call them, citing as one of his examples the benthic flatfish that live on the ocean floor, which have both eyes on the same side of the head.[27]

Although the Mendelists would turn out to be wrong—most evolutionary change does indeed occur gradually and involves natural selection—they did have a point. The real mystery of evolution is not selection, but the creation of new phenotypes. But they were born too early. They could speculate wildly, but had no way to solve the mystery, and the controversy between the two camps continued well into the twentieth century until powerful new insights would dissolve it. That process began when a long-known fact became newly appreciated: Genetic change happens not just in individuals, but in populations.

The white-bodied peppered moth is a perfectly inconspicuous insect whose white wings are sprinkled with flecks of black. Against a background of tree bark and lichen, this mottled pattern camouflages the moth against ravenous birds. In some moths, a gene affecting wing color can mutate to produce a dark-colored wing. This mutation is usually bad news for a moth, because mutant moths are no longer camouflaged, and birds can rapidly pick them off. But in nineteenth-century England the Industrial Revolution gave the dark mutant moths a much-needed break. During this time, air pollution became so severe that it wiped out most lichen and turned tree bark black. Now the dark moths were well hidden, and the white moths had turned into bird food.

If natural selection mattered, we would expect that the black moths would become more frequent over time. They would sweep through a moth population, whereas white moths would become rare. This is indeed what happened in nineteenth-century England, as the proportion of black moths in the population rose from 2 percent in 1848 to 95 percent by 1895.[28] But this information isn't nearly as important as the questions it triggers: Can we predict how rapidly they sweep through the population? Or conversely, if we have observed how fast they sweep, can we infer how strongly the dark color affects fitness, a moth's chances of remaining hidden from birds? These were quantitative, mathematical questions, new to evolutionary thinking. And they created a new quantitative discipline within biology: *population genetics.*

One of the central insights of population genetics is to view a population not just as a collection of distinct *organisms* but as a collective pool of *genes.* The genes that determine a moth's wing color, for example, have different forms—the technical term is *alleles*—responsible for light or dark wings, that occur in different proportions or frequencies in the population. Imagine that at any one time, equal numbers of both types of alleles were present in a population of organisms, and that some new factor—a new predator, or a change in pollution—allowed moths with darker wings to live longer, and so produce more offspring. Their advantage need not be huge, but even a merely 1 percent increase in the dark-winged allele, from 50 percent to 51 percent in the first generation, could accumulate over time and allow the dark-winged variants to occupy a larger and larger percentage of the population. That's how natural selection works: It changes allele frequencies, and thus the appearance of individuals over time.

This was revolutionary. The study of life, which had largely depended on the same tools since Aristotle—close observation and dissection in the field and laboratory, recorded in sketchbooks and notes—began to embrace the mathematics of differential equations and the analysis of variance. Through the minds of intellectual giants such as Sewall Wright, J. B. S. Haldane, and the statistician R. A. Fisher, population genetics developed into a theory that could answer precise, quantitative questions about natural selection. At the same time, naturalists studied the frequencies of alleles in wild populations such as that of the peppered moth, and experimentalists created evolution in action in the laboratory, by studying laboratory populations of small, rapidly breeding animals such as fruit flies. The mathematical theory was the mortar that helped join these observations into an intellectual edifice.

The new evidence from population genetics showed that variation covered a broad spectrum, with "pure" Mendelian variation at one extreme, and continuous variation at the other. Mendelian phenotypes—wing color, pea shape—are influenced by one gene with large effects. Continuously varying phenotypes like height are influenced by multiple genes, each with a tiny effect. Population genetics showed that natural selection affects both kinds of genes. But truly surprising was how powerfully selection could affect them. If a dark-wing allele decreased a moth's chance to be eaten by a few percent, it could wipe out the light-wing allele within a few dozen moth generations. And both naturalists and experimentalists found far more genes in their populations with small effects than with large ones. Mendel clearly had chosen his peas very carefully, because Mendelian traits that are influenced by a single gene comprise a tiny fraction of all traits.[29] Most evolution is gradual and does not make large jumps.[30]

By the 1930s, the concept of natural selection, the nature of inheritance, and population thinking had been synthesized into a body of knowledge known as the *modern synthesis,* named after an eponymous book by the biologist Julian Huxley.[31] Despite its name, the synthesis will soon be a century old. But unlike most centenarians, it shows no signs of senescence. Augmented by mathematical refinements and modern data, it is unbroken, and by some measures stronger than ever, playing an increasingly important role in understanding human biology—helping to reconstruct human origins, trace human migrations, and understand genetic diseases. If this edifice of knowledge were a physical building, it would rival everything architects have conceived, from the palaces of

Angkor Wat and the mausoleum of the Taj Mahal to the great Gothic cathedrals of the thirteenth century. It is a grand achievement of the human mind.

There is, however, a dirty secret behind its success. The architects of the modern synthesis focused on the genotype at the expense of the organism and its phenotype. They neglected the marvelous complexity of organisms with their trillions of cells, each inhabited by billions of molecules whose functions are themselves incredibly complex. And they neglected how all this complexity unfolds from a single fertilized cell, and how genes contribute to this unfolding. By neglecting this complexity, the architects of the modern synthesis effectively ignored its product: the organism itself. They did so knowingly, since they wanted to understand how gene frequencies change over time. In focusing on the genotype, they simplified an organism's phenotype down to simpler quantities, such as *fitness,* the average number of genes a typical individual transmits to the next generation. (Fitter organisms contribute more genes to the next generation's gene pool.) What is more, they also assumed that individual genes play a simple role in determining fitness, for example that fitness is the sum total of many small gene effects.

Don't get me wrong. It is hard to see how the modern synthesis could not have ignored the organism. The price of understanding is always abstraction, neglecting most of a staggeringly complex world to understand one tiny fragment of it. Take it from another theorist, Albert Einstein, who knew what he was talking about when he said that "everything should be made as simple as possible, but no simpler."[32] The modern synthesis was just as simple as it needed to be to answer thousands of questions about the evolution of genes and genotypes. Its very success in understanding natural selection in action was built on getting rid of organismal complexity. But whenever a theory is successful, it is also easy to forget its limitations, and this is exactly what happened in the heyday of the modern synthesis, when the grandeur of life's evolution became redefined and demoted to a "change in allele frequency within a gene pool."[33] The principal limitation—a high price to pay—was the inability to answer the second great question the *Origin* had left open: Where do innovative phenotypes come from? The modern synthesis could explain how innovations spread, but not how they originate.

To say that all evolutionists had thrown the organism under the bus, however, would be unfair to a minority of them, those who compared how the complexity of different organisms unfolds in their embryos. But these embryologists, whose forebears had helped Darwin to recognize the common ancestry of all living things, were sidelined by the modern synthesis and its advocates, who had no need for the embryo. In 1932, one year before he would win the Nobel Prize for showing how genes are organized into chromosomes, the fly geneticist Thomas Hunt Morgan would say that it does not matter much "whether you choose an ape or the foetus of an ape as the progenitor of the human race."[34]

But even though population geneticists ruled in biology's halls of power, some embryologists in the back rows kept heckling the opinion leaders, pointing out that they were ignoring the very thing they were trying to explain. Their voices got louder toward the end of the twentieth century. That's when evolutionary developmental biology, or "evo-devo," emerged as a new research discipline, one that aims to integrate embryonic development, evolution, and genetics. Evo-devo produced fantastic insights into how genes cooperate, like orchestra musicians, to make embryonic development possible.

So far, though, these insights have not yet added up to a theory rivaling the modern synthesis. And only theory can turn a heap of facts into a tower of knowledge. The culprit is once again the enormous phenotypic complexity of whole organisms. Even today, we struggle to fully understand the phenotype of even the simplest organisms, and hundreds of thousands of biologists laboring over many decades have still not fully understood how genes help shape this phenotype.[35] Where the modern synthesis has a theory without phenotypes, the embryologists have phenotypes without a theory.

Evo-devo, however, has taught us an important lesson. To understand innovability we cannot ignore the complexity of phenotypes. We must embrace it. And even though we do not yet understand all of an organism's complexity, we now understand the parts of the phenotype that ultimately bring forth all innovations. This is where the next chapters will take us.

The same century that led biology from Darwin to Mendel and the modern synthesis also gave birth to biochemistry, a science that had been conceived more than seven thousand years earlier, when humans started to produce beer and wine. The mechanism by which yeasts transform sugar into ethanol remained mysterious, however, until Louis Pasteur showed, three years before Darwin's *Origin,* that living organisms cause fermentation. And even that truth was toppled a few decades later, when Eduard Buchner proved in 1897 that fermentation does not require living organisms, because yeast extracts containing no living cells can ferment sugar. His discovery helped dispel vitalism, the notion that life required an enigmatic vital force and obeys laws different from those of the inanimate world.

To teach us that life is based on prosaic chemistry is important, but Buchner is even better remembered as a pioneer in the discovery of enzymes, those gigantic protein molecules consisting of dozens to thousands of amino acids.[36] They can speed up chemical reactions that cleave, join, or rearrange atoms up to a billionfold. Biochemistry honors Buchner to this day by using his naming system for enzymes, adding the suffix *-ase* to the chemical reaction they catalyze. An enzyme that can process the sugar sucrose would be *sucrase,* one processing lactose would be *lactase,* and so on.

His discoveries also spun off another branch of biochemistry. It focused not on enzymes but on the reactions they catalyzed, and would unveil a new chemical world, that of metabolism with its bewildering complexity. Broadly speaking, an organism's metabolism—the word itself comes from the Greek for "change"—comprises two sorts of chemical transformations. The first kind cleaves energy-rich molecules such as the sugar glucose to extract energy from them. The second uses this energy to transform nutrient molecules into a cell's own molecular building blocks, which comprise dozens of molecules like the amino acids in proteins. Along the way, a metabolism must also manage a body's waste, disarming toxic molecules into harmless ones. Taken together, these tasks are complex and require more than a thousand chemical reactions—and the enzymes that catalyze them—in order to build and maintain our bodies.[37]

The discovery that protein enzymes help build our phenotype is a monumental insight of twentieth-century biochemistry. (It also led to a key insight about life's creativity: Even the largest changes in an organism result from alterations in individual molecules.) But this discovery was dwarfed by an even greater one: the chemical structure of our genes.

Its story also begins at Darwin's time, in 1869, the same year as the fifth edition of *The Origin of Species.*[38] That is when the Swiss chemist Friedrich Miescher first identified a new mysterious substance, different from protein.[39] He called it *Nuklein,* but its chemical structure would not become clear until decades later. Not until 1910 would we know that the substance—by then renamed deoxyribonucleic acid (DNA)—contains the four bases adenine (A), cytosine (C), guanine (G), and thymine (T), molecules that we now call the four letters of the DNA alphabet. And it would be 1944 before biologists realized that DNA is the stuff of inheritance. In that year Oswald

Avery showed that DNA from a disease-causing strain of the bacterium *Streptococcus pneumoniae* helps another, harmless strain kill mice.[40]

Less than a decade later, James Watson and Francis Crick would reveal that DNA is a supremely beautiful molecule. Its two strands form the famed double helix, a twisted ladder in which two paired bases from opposite strands make up each rung. In each rung, two bases are always paired, A always with T, and C always with G. This structure also suggests how DNA could be copied, and thus how inheritance works at the level of molecules.[41] Genes had turned out to be so much more than Johannsen had thought.

It had taken seventy years to get from Muybridge's zoopraxiscope to color television—to get from recording individual black-and-white images on silver plates to encoding color images as electric signals, transmitting them wirelessly, and displaying them on cathode-ray tubes.

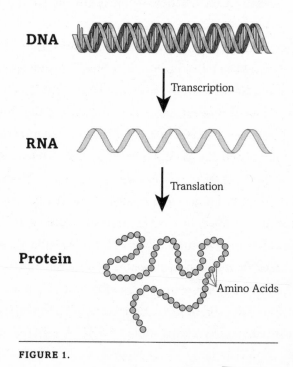

FIGURE 1.

During the same seventy years, biology had also progressed dramatically and embraced new discoveries just as enthusiastically. It had married the mathematics of population genetics and birthed the modern synthesis. It had revealed the function of enzymes and discovered the structure of DNA (brought to us at about the same time as color television). It had incorporated the knowledge of chemistry that would become essential to understanding the origins of innovation. It wasn't there yet. But it was getting closer.

Watson and Crick's discovery rang in the age of *molecular biology*. Within the next twelve years, biologists would learn that DNA is transcribed into the closely related ribonucleic acid (RNA), which is then translated, three nucleotide letters at a time, into a protein string of amino acids (figure 1). This translation follows a genetic code in which most of the sixty-four possible three-letter words encode a single amino acid. Only a few words are set aside to signal the beginning and the end of a protein string.

Knowing the DNA letter sequence of a gene, a child could predict the amino acid sequence of a protein. But this is where simplicity ends. Proteins fold into intricate three-dimensional shapes that wobble and vibrate. To understand how they perform their tasks, such as to accelerate chemical reactions, both the shapes and their vibrations need to be known. And to this day we are unable to predict either one from the underlying amino acid string, so complex and subtle are the rules underlying this folding. To be sure, experiments to identify protein folds were already under way in the 1950s, beginning with the oxygen-storing globin proteins of our blood and muscles.[42] But these experiments were laborious and could take years. Whereas finding the amino acid string encoded by a DNA letter sequence is as easy as looking up a word in a dictionary, predicting a protein fold is much harder—a bit like translating a poem by Yeats into Chinese.

This is not good news for anyone hoping to understand where

innovative phenotypes come from. Understanding an organism's phenotypes—any of its aspects, whether the color of a wing, the acuity of an eye, or the strength of a bone—comes down to understanding the molecules that build a body, the smallest building blocks of the phenotype. If we cannot predict their shape, it is impossible to travel the road from the genotype all the way to the phenotype. But that road is where nature innovates. Without understanding its twists and turns, its speed limits and traffic signs, we know little more about innovability than Darwin.

And it gets worse, because proteins don't operate on their own. They cooperate like worker bees in solving a complex task. Take the protein hormone insulin, a messenger molecule produced by the pancreas that commands your liver cells to absorb and process glucose. Insulin cannot enter the liver directly. Instead, it binds to a protein on a liver cell's surface, the insulin receptor. In response, this receptor modifies another protein inside the cell, which starts a chain of handshakes between further proteins that, eventually, turn on the genes needed to process glucose. At every moment of our existence, thousands of such molecular signals crisscross our body and are processed inside cells. Since Watson and Crick's discovery, molecular biologists have increasingly studied processes like these. Pulling on a few loose strands, they have unearthed the molecular webs that allow us to eat, move, see, hear, think, taste, sleep, and do just about everything else we do.

But we got more than we asked for. Thousands of man-years have already been poured into this endeavor, and the end is not near. To the contrary, the more we learn, the more strands of this web become evident, the more complex and tangled it seems. The road from genotype to phenotype extends to the horizon and beyond.

Throughout the twentieth century, many evolutionary biologists were undistracted by all this complexity. Basking in the glow of the modern synthesis, they were blissfully focused on the genotype. And this focus became even greater after Watson and Crick's work had stirred the ocean of our ignorance, and after new technology to read the letter sequence of DNA molecules had been developed. This technology spawned a new research field known as *molecular evolutionary biology*, whose subject was variation in amino acid and DNA strings. The earliest incarnation of this technology was about as inefficient as Muybridge's zoopraxiscope—a year's work would reveal no more than a few hundred letters. By the mid-1980s, however, its efficiency had increased more than tenfold, enough to read short sequences of DNA from multiple individuals in a population.[43]

When molecular evolutionists took advantage of this technology, they discovered something nobody had expected: enormous amounts of genetic variation, everywhere, even in organisms that had not changed for many millennia.

One early molecular evolution study focused on *alcoholdehydrogenase*, an enzyme that helps detoxify ethanol. We have a gene for it, and so do fruit flies. No one knows whether they get as high on fermented fruit as any Skid Row wino, but they certainly are attracted to it, and they need this enzyme to prevent alcohol poisoning. In 1983, Martin Kreitman from Harvard University found that the DNA from a small sample of fruit flies contained more than forty-three different DNA text variants in this gene.[44] Similar variants occur in humans. One of them causes a form of alcohol intolerance where blotches erupt on the faces and bodies of sensitive individuals, a condition so widespread among people with Asian ancestry that it is known as "Asian flush."[45]

But what Kreitman did *not* find in the alcoholdehydrogenase gene was even more telling. Most of the mutations in this gene were *silent*. They changed the DNA sequence, but not the amino acid sequence of alcoholdehydrogenase. This is possible because the genetic code is redundant, because more than one three-letter word can encode the same amino acid. And it was surprising. Even with a redundant code, there should have been many more amino-acid-changing mutations, because mutations tend to sprinkle genes randomly with letter changes. Something had happened to these mutations.

The something was natural selection. Because these changes impaired the enzyme, natural selection had weeded them out long before Kreitman got to see them.

Kreitman's discovery and others like it illustrate a fact that is easily overlooked: The revolutions in evolutionary thought are different from other scientific revolutions. Whereas the revolution of quantum physics in the early twentieth century, for example, gave rise to a worldview incompatible with that of classical physics, revolutions in evolutionary biology leave core elements of previous theories intact.[46] Rather than overturning the past, they deepen and sharpen it. They add layers of clarity and resolution, as well as new dimensions. The film *Seabiscuit* added color, music, dialogue, and the sound of hoofbeats to the first recording of Sallie Gardner's ride, but it didn't invalidate Muybridge's revelation about the nature of galloping. Where Darwin used the natural world to infer the power of selection, the modern synthesis could see it in the ebb and flow of gene frequencies, and molecular evolutionists found it in DNA signatures, such as the excess of silent mutations. In doing so, they dissolved a fog of confusion that Darwin left behind. (*Some* of the fog, because the molecular revolution taught us more about genotypic than phenotypic change, the heart of the origination problem.)

The amount of variation Kreitman found in the alcoholdehydrogenase gene is not unusual. Animal and plant populations are chock-full of genetic variation. Genetic variants even occur in populations of living fossils whose phenotypes have not changed for many millions of years, such as the coelacanth, a strange fish thought to be extinct until a live specimen was found in 1939.[47] Their abundance raised questions that occupy molecular evolutionists to this day. Do most of them matter for phenotypic evolution? Are they necessary or irrelevant for life's innovations? Their mere existence underlines how hard it is to understand phenotypic innovation and how it emerges from genetic change.

The ability to read a thousand letters of DNA text was still impressive in the 1980s. But a thousand letters are nothing compared to an organism's genome, the totality of its DNA. Human DNA is three billion letters long, ten times longer than the *Encyclopædia Britannica*. Every single one of the trillions of cells in our body contains a copy of it, packed into our forty-six chromosomes. Even the DNA of a bacterium like *Escherichia coli* has four and a half million letters, more than *War and Peace,* one of the longest novels ever written. DNA sequencing technology needed to get much better to read just the genome of a single individual, let alone to catalog variation in an entire population.[48] The impetus to develop this technology would come from the Human Genome Project, one of the largest international research collaborations ever, initiated in 1990 and spearheaded by the U.S. National Institutes of Health. This is not a coincidence, for the project aimed at understanding genes that cause disease, a special kind of new phenotype. Fierce competition to this publicly funded effort arose in 1998 from the company Celera Genomics and its founder, the biologist and entrepreneur Craig Venter. They managed to sequence the genome at a tenth of the cost, and crossed the finish line simultaneously with the publicly funded project in 2000, when a first draft human genome was published.[49]

The human genome is another major milestone of biology that revealed a host of genetic information, how many genes we have, what proteins they encode, and so on. "The blueprint of life" is what President Bill Clinton called it in his 2000 State of the Union address. But if so, it is a very odd blueprint, one that we cannot use to build what it depicts, or even to guide a repairman to fix a problem. Because thus far, the genome has guarded the secrets of our phenotype well. Many had hoped, for example, that the genome would give us yes-or-no answers to the question of whether a person would get a genetic disease. But here is what Craig Venter himself had to say about our ability to predict disease in a 2010 interview with the German magazine *Der Spiegel*:

> We have, in truth, learned nothing from the genome other than probabilities. How does a 1 or 3 percent increased risk for something translate into the clinic? It is useless information.[50]

This assessment is stark, but it holds a grain of truth. You guessed the reason: The relationship between genotype and phenotype is complex beyond imagination. The Human Genome Project is only a mile marker on the journey from genotype to phenotype. It's not anywhere close to the end of the road.

Whatever its limitations, the genome project had many other benefits. One of them was that it whipped DNA sequencing technologies to blazing speeds. While in the year 2000 one person could read up to a million DNA letters in twenty-four hours, sequencing machines available in 2008 could already read up to a billion letters within the same span, and technologies have gotten much faster since then. As of this writing, full sequencing of a human genome costs little more than $1,000, and by the time you read this, the cost may have dropped to pennies. These technologies allow us to study genomic variation in large human populations and in many other organisms. They have transformed population genetics into *population genomics*.

Population genomics is the end of the road for studying genotypes. The same cannot be said for the phenotype. The molecular biology work that began in the mid-1950s to unravel the functions of proteins and their interactions continues undiminished. But in the 1990s it had to take a new tack to progress further. For processes like insulin signaling, it had previously identified key genes, the proteins they encoded, what these proteins do, and which of them interact.[51] All this information is like a who-is-who and a who-knows-whom of the cell. In the 1990s it became clear that such a catalog would fall short of predicting phenotypes, such as whether a person would develop diabetes. It fails to capture many subtleties that matter, the number of protein molecules involved, how firm their handshakes are, and so on. Dozens of different *kinds* of molecules contribute to diabetes, each contributing only a few percent to increased disease risk, but each conspiring with multiple others in subtle—and ill-understood—ways to cause disease. For these reasons it would not get us anywhere to merely list these molecules and their properties. We need to understand exactly how these molecular parts cooperate to form a whole phenotype.

The only tools that could offer this integration are mathematical. Equations can encapsulate a wealth of experimental data and describe how the concentrations and activities of molecules change over time.[52] And these activities are key to understanding phenotypes. For example, in type 2 diabetes the body shows insulin resistance, a phenotype different from that of a healthy individual: The pancreas releases insulin, but the liver reacts sluggishly. Somewhere along the signaling chain starting at the insulin receptor, the handshake of signaling molecules has become too weak (or too firm).[53] And this change percolates down the signaling chain to cause disease and suffering. Only the rigorous quantification of mathematics can help us understand such subtleties. No mere catalog of molecules could achieve that.

There is only one catch with the equations that can describe molecular phenotypes: They are not simple. They have many variables—molecules and their interactions—distilled from decades of experiments. They cannot be solved with pencil and paper. They are beyond the capabilities of even today's most skilled mathematicians. Their solution requires computers.

Computers have become as essential to twenty-first-century biology as digital cameras have become to photography. Computers do more than just run scientific equipment—from ultracold freezers to espresso machines—they are now instruments in their own right. Like the microscopes of the seventeenth century, they allow us to travel a new world, one so small that the most powerful imaging technologies, including electron microscopes, cannot resolve it: the world of molecules. Indeed, computers *are* the microscopes of the twenty-first century. They help us understand molecular webs that Darwin did not even know existed.

This centrality of computing is new, because for much of its history, biology was limited by data. Early explorers had to voyage for years to discover new life forms in faraway lands. Even early in the molecular era isolating a single gene could be years of work. No longer. Thanks to ever-accelerating technology, thousands of ever-growing databases are overflowing with biological information, not just about genes and genomes, but also about the millions of molecular parts living things harbor, about what these parts do and which other parts they interact with. Every year now, gigabytes and terabytes of new data enter these databases. A new generation of scientists—computational biologists—uses only knowledge gathered by others, and no longer experiments with living organisms. Biologists are being transformed into information scientists, with access to nearly limitless data. The limits exist in our imagination, and in our skills to detect laws of nature in that data.

These skills will surely be challenged, because the puzzle of how new phenotypes come into being has stymied science for more than a century. It's one thing to recognize that phenotypes are like enormous pointillist paintings, created one molecular change at a time. It's another to use that insight to understand how those paintings are actually created. The challenge is daunting, even on the smallest scale of proteins like the alcoholdehydrogenase that stands between you and Death by Happy Hour, since there are more ways to string amino acids together than there are hydrogen atoms in the entire universe. Referring to random change, recited like a mantra since Darwin's time, as a source of all innovation is about as helpful as Anaximander's argument that humans originated inside fish. It sweeps our ignorance under the rug by giving it a different name. This doesn't mean that mutations don't matter, or that natural selection isn't absolutely necessary.[54] But given the staggering odds, selection is not enough. We need a principle that accelerates innovation.

Until a few years ago this principle was not merely unknown but beyond reach, and this book could not have been written. Because life is built of molecules, we need to understand molecules to understand innovation: not only the genotype embodied in DNA, but how this genotype helps build a phenotype. And a phenotype like that of a human body is not just a string of DNA. It is a hierarchy of being that descends from the visible organism, its tissues and cells, to the molecular webs formed by metabolic molecules, signaling molecules, and many others, extending down to the level of individual proteins. New phenotypes can originate at each level. A mere thirty years ago, we knew little of this staggering complexity.

And if *we* knew little, just imagine how much less Darwin knew. The list of things that he didn't know is practically an encyclopedia of modern biology. He wasn't just ignorant of how phenotypes were inherited. He also had no knowledge, in those pre-Mendelian days, of genes, to say nothing of DNA and the genetic code. He also knew nothing of population genetics and little of developmental biology—he was oblivious to how molecules build bodies. He had no inkling of life's true complexity (and many after him thought they could safely neglect it). But to crack the secret of innovation, we need to embrace it.

The time-honored way to study life's complexity is to focus on one or a few genotypes and their phenotype. This is how early geneticists found many genes in the first place—by tracking a phenotypic change back to its origin in a mutated gene. Later in the genome era, the same idea worked well to find out what a stretch of DNA does: Mutate it and see what happens to the phenotype. These strategies led to striking discoveries, mutations in genes that create flies with two pairs of wings instead of one, plants with transformed leaves, microbes able to survive on new foods. They created many examples of mutant genotypes and strangely altered phenotypes.

The problem is that examples are not enough. Explorers cannot chart a newly discovered continent by making a single landfall and taking a walk on the beach. They need to circumnavigate it to draw its contours. They have to sail into its interior from its river deltas. And they need to traverse its mountain ranges, deserts, and jungles. We need to do just that to draw the elusive maps of life's creativity—the genotype-phenotype maps that chart each change in a genotype and how it affects the phenotype. We need genotype-phenotype maps to complete Darwin's job.[55]

Even with the best technologies, these maps are not easy to draw. For a high-resolution map, we would need to understand the intricately folded phenotypes of more than 10^{130} different amino acid strings, and that's before adding any of the higher layers of a phenotype, brought forth by thousands of genes and proteins. In other words, drawing a high-resolution map is not just hard but impossible. Luckily, though, we do not have to map every grain of sand in this new continent. If we care just about its topographical features, we can get away with studying fewer genotypes. But we still need to examine thousands to millions of them. And therefore we need to choose carefully which of the myriad aspects of phenotype we study. We need to choose those that are important for innovation in life's history, and where existing information or predictive technology is sufficient to draw the map.

In these maps, Platonic essentialism is making a comeback, after decades in which it served as the antihero of evolutionism.[56] The essentialism of the twenty-first century, though, is much richer than Plato's world of simple geometric shapes. It reveals a world full of meaning, compatible with Darwinism but going far beyond it, that is key to understanding how nature creates. This world is inaccessible to our naked eyes, just like the question of whether all four legs of a galloping Sallie Gardner leave the ground, but we can explore it with the best technologies currently available.

These technologies have helped us reveal a Platonic world of crystalline splendor, the foundation of life's innovability, which began with life's very origin some four billion years ago.

CHAPTER TWO

The Origin of Innovation

H ere is an amazing experiment you can try at home. Put wheat in a container and seal the opening with dirty underwear. Wait twenty-one days, and mice will emerge. Not just newborn mice, but grown adult mice. At least that's what the seventeenth-century physician and chemist Jan Baptista van Helmont reported.[1] (He also revealed that scorpions would emerge from basil placed between two bricks and warmed by sunlight.)

Van Helmont wasn't the first to postulate the doctrine of spontaneous generation, which dates back at least to Aristotle, though he was among the last. Today, any scientist reporting that wheat and underwear conspire to create new life would be forever branded as a crackpot, but Van Helmont's sloppy experiment did not cause much of a stir, and he died a respected man in 1644. Spontaneous generation was so widely accepted in his time that his experiments just proved the obvious.

A few decades after Van Helmont's death, the Italian physician Francesco Redi showed us how experiments like this should be done.[2] Dump meat in a jar and in good time it will be crawling with maggots. But not spontaneously created ones: When Redi covered the jar with muslin, no maggots emerged, because flies could no longer deposit their eggs in the meat.

Redi helped speed the decline of spontaneous creation. So did the seventeenth-century Dutch fabric merchant and lens grinder Antonie van Leeuwenhoek, whose microscopes opened the door to the world of microbes. For a time, microbes, being so much smaller than visible life, offered refuge to the remaining advocates of spontaneous generation. These were people like the Scottish priest John Needham, who argued in the mid-eighteenth century that decaying organic matter created microbes.[3] Another century later Louis Pasteur would show that Needham had it backward: Microbes cause the decay of organic matter, not the other way around. Pasteur hammered the last nail into the coffin of spontaneous creation when he sterilized a nutrient broth and the air around it and showed that it remained lifeless.[4]

Pasteur could show that spontaneous generation didn't exist, but he and his contemporaries could not have known why: The origin of life is a problem for chemists, not biologists. And chemists in the nineteenth century suffered from the same disease as the Mendelists who tried to understand new variation in the early twentieth century: They were born too early. Dmitri Mendeleev had barely worked out the periodic table of the elements, and the chemistry of life was a big blank spot. Chemistry in general took a long time to become a respectable science in its own right, perhaps because of its deep roots in alchemy. Well into the twentieth century, after his first wife had run off with a chemist, the Nobel Prize–winning quantum physicist Wolfgang Pauli would remark to a friend that "had she taken a bullfighter I would have understood, but an ordinary chemist . . ."[5]

A century later we know that the overwhelming obstacle facing spontaneous generation is probability, or rather improbability, resulting from life's enormously complex phenotypes. If even a single protein, a single specific sequence of amino acids, could not have emerged spontaneously, how much less so could a bacterium like *E. coli* with millions of proteins and other complex molecules? Modern biochemistry allows us to estimate the odds, and they demolish the spontaneous creation of complex organisms.

This does not mean that spontaneous creation did not occur in life's early history. A natural origin of life even requires it, but in a much humbler form than a modern cell or even a modern protein. Earth's first life form was far more like an oxcart than a Ferrari. In fact, it was a lot more like a wheel than an oxcart. And even this wheel was not created in one giant leap, but in many modest steps. Although the muck of deep time has eroded their footprints, chemists have reconstituted some of these steps, which are the subject of this chapter. They not only illustrate how it could have happened but prove an even more important point: Even before life itself arose, nature's creativity used the same principles it uses today. Then and now, the new and improved arrives through new chemical reactions and molecules.

The Hadean Eon, which marks the beginning of earth's geological history more than four billion years ago, is aptly named after the Greek underworld, because the early earth was a hellish place. It began with a surface of liquid magma surrounded by an atmosphere of vaporized rock.[6] And even after the surface had congealed into a solid crust, Mother Earth was not an inviting place. Had you visited the Hadean earth from outer space, you would have seen a tortured skin pockmarked with countless volcanoes, steamed by scalding rains that poured into the primordial oceans. Only the enormous pressure of the atmosphere—much denser than today—prevented these oceans from boiling away. Needless to say, breathing this atmosphere would have felled you instantaneously, noxious as it was from deadly amounts of carbon dioxide and hydrogen.[7] Ducking for cover might also have been smart, for multiple giant asteroids tore into early earth during a period called the Late Heavy Bombardment. You can still shudder at their scars, giant craters visible nightly on the moon, even though the churning continents here on earth have erased most visible traces of these ancient cataclysms. We know their age—and that of earth itself—from ancient rocks that contain slowly ticking chemical clocks, materials like uranium, whose radioactive decay marks the passing eons.

Most remarkable about this period is the speed with which life got going once the worst was over, just about 3.8 billion years ago. A mere few hundred million years later—less than 10 percent of earth's age today—the first fossilized microbes appear.[8] Even closer to the magic boundary of 3.8 billion years ago, telltale traces of an ancient metabolism in the form of light isotopes of carbon appear in rocks from West Greenland.[9] Life wasted no time, and appeared almost as soon as it *could* appear. This tells us that life's origin and the innovations behind it might not be that hard to come by. And that innovability is as old as life itself.

Life's early appearance on earth demands a theory of its chemical origin. Among the earliest ones is the "primordial soup" theory, usually credited to Alexander Oparin and J. B. S. Haldane, the Haldane of modern synthesis fame, who wrote about it in the 1920s.[10] Remarkably, however, the ever-prescient Charles Darwin had this idea half a century before them. In an 1871 letter to his friend Joseph Dalton Hooker, he speculates, "If (and oh what a big if) we could conceive in some warm little pond with all sorts of ammonia phosphoric salts, light, heat, electricity etc present, that a protein compound was chemically formed, ready to undergo still more complex changes." And in the same breath Darwin gives us a good reason why we might look in vain for such a warm little pond today: Its content would be instantly "absorbed or devoured" by today's organisms.[11]

The primordial soup remained speculation for decades, until 1952, when it received a huge boost from Stanley Miller, a graduate student in the laboratory of Nobel laureate Harold Urey at the University of Chicago. Based on an informed guess at the composition of the gases that were present in the early atmosphere, Miller sealed these gases in a container, showered them with electric sparks to simulate primordial lightning, and washed the mixture in a rainfall of condensing water. After mere days, many organic molecules—those normally created by living organisms—had appeared in Miller's miniature world. This was a monumental discovery, for it showed how organic molecules could have emerged from inorganic matter during the turbulent youth of our planet.[12] And Miller's primordial ocean produced not just any organic molecules. It created amino acids such as glycine and alanine, basic building blocks of modern proteins.[13] Later experiments produced many other of life's construction materials, including sugars and parts of DNA.[14] But even more important was that Miller's experiments moved life's origin from philosophical speculation to the realm of hard, experimental science.

In September 1969, the world learned something that Miller had not known in 1952: Life's molecules can emerge in environments even more hostile than that of the early earth. That September an exploding fireball briefly created a second sun in the sky over Murchison, an Australian town of a few hundred souls some one hundred miles north of Melbourne. After fracturing, the meteorite left a trail of smoke and smaller fragments, the largest of which fell harmlessly into a barn. This cosmic accident happened two months after man had first walked on the moon, at a time when scientists were itching to study extraterrestrial rocks.

As they scratched this itch, they found that the Murchison meteorite ferried a most unusual cargo. As old as the earth itself, it had wandered through outer space for eons, yet it contained several of the amino acid building blocks of proteins, as well as purines and pyrimidines, which are important DNA building blocks. Later work used twenty-first-century spectroscopy to show that it harbored more than ten thousand different kinds of organic molecules, although many of them in exceedingly small quantities.[15]

The Murchison meteorite, it is important to know, is not a freak of nature. Similar meteorites have landed on earth, and countless other rocks ferry organic cargoes through the heavens.[16] Fortunately, we no longer need to wait until another one of them drops. Because molecules in the universe absorb or emit radiation that reveals their structure, the hypersensitive ears of radio telescopes can distinguish hundreds of different organic molecules whose voices whisper to us in multitudes from clouds of interstellar gas. Actually, they shout, since three-quarters of the molecules in these interstellar clouds are organic, and include key constituents of life, such as the amino acid glycine.[17] Incidentally, the single most abundant three-atom molecule in interstellar clouds is water, another blow to the notion that we and our planet are oh so very special.

Life's simpler building blocks are so prevalent in the universe that molecules from space may have seeded life on earth itself. Meteorites and comets, especially those that bombarded the early earth, discharged ten times more water than currently fills all of the earth's oceans, and a thousand times more gases than its present-day atmosphere.[18] What is more, they also delivered the rich buffet of organic molecules we find in interstellar space, and in a staggering number of servings. At least ten trillion tons of organic carbon, and perhaps a hundred times as much, have entered our atmosphere from outer space.[19] That is at least ten times more than all the carbon that circulates in living cells today. Especially important is the dust trailing behind comets that pass our orbit. Unlike large meteorites whose white-hot temperatures destroy some of their organic cargo during their explosive landing, cometary dust merely blankets the earth in an invisible but unceasing rain of life's seeds.[20] Perhaps we really are made of stardust.

We may never know whether most of life's molecules were created in outer space or on earth. But regardless, these observations contain some simple and important lessons. The first is that life's molecules emerge spontaneously in the right environment. The second is that this environment need not be, like Darwin's warm pond, a nearby and very special place in the universe. It could be light-years away or as ubiquitous as interstellar gas.

The third is a lesson about innovation—I already mentioned it—that is still valid today: Innovation revolves around new molecules and the reactions that create them. To understand innovability, we need to understand the origins of these molecules.

The molecules of life are not yet life itself, any more than a pile of bricks and lumber is a mansion. At a minimum, life needs a metabolism, a network of chemical reactions that harvests energy and combines chemical elements into life's molecular building blocks. Life also needs the ability to make more of itself—to *replicate*—and pass its accomplishments on to future generations as heritable traits. Without offspring that resemble their parents, Darwinian evolution would be unworkable, natural selection impossible.

This doesn't mean that metabolism and replication must have appeared simultaneously. Even today they do not always occur together. Viruses replicate but have no metabolism of their own, hijacking instead the metabolic machinery of their host cells. But true life requires both metabolism and replication, and that requirement raises the earliest chicken-and-egg question ever: Which came first?

Perhaps seduced by the ethereal beauty of the DNA double helix, mainstream science held for years that replication came first. But to explain its origins is a tall order, because modern replication is extremely complex. What's more, the nucleotide letters of DNA do not replicate themselves. Rather, they only carry information. They are transcribed into RNA, which is translated into proteins (figure 1), and these proteins are tasked with just about everything else, including transcription and replication. Because no one protein masters all the necessary skills, dozens of proteins share these tasks, and each protein has a precisely specified amino acid sequence. This sophisticated division of labor leads to another chicken-and-egg problem, this one about whether proteins or nucleic acids—the collective term for DNA and RNA—appeared first. To have both emerge simultaneously would be asking too much—we would be back to odds-defying spontaneous creation. But if the first life consisted of a single replicator, this molecular Adam (or Eve) would have to be spectacularly versatile, able to both carry information *and* copy itself.

In their 1953 discovery of the double helix, Watson and Crick had already recognized that the key to replication was the pairing of complementary DNA bases—G with C and A with T—which glues the two strands of the double helix together. In their words, it "immediately suggests a possible copying mechanism for the genetic material."[21] That very mechanism excludes proteins as the first replicators, because their amino acid parts cannot transmit information in this way. They lack the simple complementarity that allows two DNA strands to build the twisted ladder of the double helix.

So proteins are poor replicators. But nucleic acids seemed to be just as poor at everything else. Could they do what proteins excel at? Could they catalyze their own replication? Could they catalyze anything at all? The very purpose and structure of DNA made that seem unlikely. DNA's primary task is to do as little as possible except to store information, inertly, faithfully, generation after generation after generation.[22] For more than half a century after the discovery of enzymes, most scientists thought that only proteins, not nucleic acids, could catalyze chemical reactions.

And so the stuff of the first replicator remained mysterious. Until 1982, that is, when the chemists Thomas Cech and Sidney Altman transformed RNA from an ugly duckling into a white swan.[23] RNA had been the stepchild of molecular biology, largely a messenger ferrying information from DNA to the ribosome, the hugely complex molecular machine that synthesizes proteins.[24] But these two chemists jolted science with the discovery that RNA can catalyze chemical reactions all by itself.

The knowledge that RNA could perform the job of proteins catalyzed—pardon the pun—many other discoveries. Before long, biologists realized that RNA had an ancient history, even older than that of proteins and DNA, and that RNA had ruled over a sunken world of early life.[25] Unlike the fabled Atlantis, however, this world has left many traces. One of them is that RNA remains a key molecule in today's command centers of life. For example, in the ribosome machine, which contains dozens of proteins and also a few RNA molecules, the RNA—not the proteins—catalyzes the concatenation of amino acids into proteins, including those that make the ribosome's own proteins.[26]

RNA may once have both carried life's information and helped catalyze its own replication, but we are none the wiser how it got there. To find that *first* innovation, the origin of life itself, it would help if we could construct a simple molecule that replicates itself. This molecule would be an RNA *replicase,* an enzyme that catalyzes the replication of RNA.[27]

Some of today's finest chemists are on a quest to find this single replicase. Their best efforts so far created a 189-letter-long RNA string with *some* talent as a copyist—it can't actually copy itself, but only a shorter molecule template of about fourteen letters.[28] This tells us that RNA-based replication might just work, if several obstacles can be surmounted. One of them comes from the very feature that makes nucleic acids replicable: base pairing. Complementary bases stick together, which means that a parent molecule and the complementary copy strung together by a replicase would anneal into a double-stranded RNA molecule like the familiar double-stranded DNA. To make another copy from them, the two strands would have to be separated, so their information can be read. But as soon as you—or a replicase—pulled them apart, their sticky bases would anneal again like two ribbons of Scotch tape. The very same feature that allows replication is also its worst enemy.

Another problem is that the first replicase would have needed to be unfathomably accurate, because a sloppy replicase would trigger an *error catastrophe,* a process discovered and baptized by the Nobel Prize–winning chemist Manfred Eigen.[29]

To understand error catastrophes, it is helpful to think of how medieval monks copied sacred texts, transcribing letter by letter by tedious letter. If one monk misread a letter, the error might be inherited in the copy. A second scribe might propagate the error, introduce errors of his own, and so on, until monk by monk, over generations and centuries, the text might slowly erode to a meaningless jumble of letters. An RNA replicase, whose one and only ability is encoded in a molecular text, faces a similar problem, but with an added twist: In an RNA world, the monk and the text he copies are one and the same. The replicase is a book that transcribes itself, and its errors erode not only the text itself but also its own ability to copy. Subsequent generations of monks become ever more error-prone.

Only a replicase that creates mostly error-free copies of itself can preserve the information in its letter sequence, which encodes the very ability to replicate itself. But if it is too sloppy, most of its copies will be inferior replicases—slower, perhaps, not as faithful—and will degrade over time into useless molecules from which the original information is erased. In 1971, four years after winning the Nobel, Manfred Eigen calculated the accuracy required to evade this error catastrophe. The longer a replicase is, he found, the more accurate it needs to be. As a rule of thumb, a replicase with fifty nucleotides would have to misread fewer than one in fifty nucleotides, a replicase with one hundred nucleotides would need to misread fewer than one in a hundred nucleotides, and so on.[30] The current best candidate of 189 letters misreads several times as many.[31] Even if it could replicate itself, it would gallop straight over the cliff of the error catastrophe.

Fortunately, life does much better than that today. The DNA-copying machinery of protein enzymes misreads fewer than one in a million letters.[32] But its accuracy comes at a price: complexity. The machinery includes highly specialized proteins that proofread and correct the errors of other proteins, as if every text were copied by groups of monks looking over one another's shoulders. These proteins are encoded by long genes, much longer than any primordial RNA replicase could have been. And to preserve information over time, the replication they supervise needs to be highly accurate. You may already see the next chicken-and-egg problem lurking here. It is also known as Eigen's paradox: Faithful replication needs long and complex molecules, but long molecules require faithful replication. To this day, nature has not shown us an exit from this labyrinth, but as we shall see in chapter 6, a principle of innovability found in today's life provides a clue.

The unwelcome stickiness of RNA and the catastrophic Eigen's paradox are already daunting obstacles to the idea that replicators were life's first innovations. But they are mere foothills to the Himalayas of a third problem: finding a sufficiently rich supply of raw materials—energy-rich molecules that also contain all the needed chemical elements, including carbon, nitrogen, and hydrogen. Examples include the energy-rich precursor nucleotides of DNA's letters, of which modern replication proteins consume nearly a thousand every second when they copy DNA.[33] And even if an early replicase were much slower and more inefficient—perhaps copying one letter per second, replicating itself in some three minutes—the demand for raw materials wouldn't disappear.[34] Since each copy would itself be a replicase, both the number of copies and the ability to make even more copies would increase in lockstep, which means that even a glacially slow pace of replication by modern standards would still result in an exponential population explosion and a huge demand on nucleotide raw materials. After the first six hours, this population would already have devoured 1 ton, after a day 2.5 tons, and after a week no fewer than 800,000 tons of nucleotide raw materials.

Once life has arrived, it quickly multiplies into an army of molecules that is voraciously hungry for a continuous supply of energy-rich materials. Like any military expedition, it would quickly collapse without such a supply chain. What is more, absent a steady food supply, Darwinian evolution and natural selection cannot work. Their power unfolds over many generations and thus needs replication—a lot of it. In addition, it does not help that replicases, like soldiers, are mortal.[35] Over time, they decay through random collisions with other molecules. Starving, they would decay faster than they could make copies. Life's campaign to conquer the planet would fizzle out like a wet match, moments after it had ignited.

The supply chains created by Miller's experiments and by interstellar chemistry are not strong enough to sustain this army. Although they create molecules like amino acids that life craves as raw materials, they do not create enough of them to feed life continuously. Miller's experiments take days to produce a few milligrams of organic molecules from about a kilogram of carbon.[36] And while meteorites may help import megatons of organic carbon given enough time, the first replicators would starve without having food in enormous quantities, at the same time, and in the same place. Depending on meteorites to sustain life is like relying on a manure-carrying truck to crash into your backyard garden every few days.

All this leaves a gnawing suspicion that the replicator-first idea puts the cart before the horse. Seduced by the beauty of the double helix, its advocates have dreamed up a gleaming and sophisticated automobile factory—before reliable parts suppliers existed. The factory's high-throughput assembly line is useless until wheels, axles, transmissions, and engines can be built in quantities. And if the supply's trickle is so slow that only one car can be built every few years, then decay and eventual bankruptcy become inevitable. The obvious alternative is that before a single self-replicating molecule could emerge, the supply network had to be in place, a network of chemical reactions that could produce life's raw materials.

In other words, life started not with a replicator, but with a metabolism.[37]

When the right molecules come close enough, the chemical reactions needed to produce energy and life's building blocks proceed—eventually. But *eventually* can be a long time, a *very* long time. Some of life's chemical reactions would take thousands of years to proceed without help. For this reason, metabolism needs catalysts, molecules whose main job is to speed up chemical reactions. Catalysts have a remarkable feature: Driven by heat—the incessant bouncing and vibrating of atoms and molecules—they can arrange other molecules such that their atomic parts come in contact and react, but they themselves stay above the fray and are not eaten up by the reaction. Catalysts are accelerants for the fire of metabolism. Their main job is to lower the activation energy for a particular chemical reaction, and accelerate it by several orders of magnitude. The catalysts of modern metabolism are protein enzymes, extremely efficient and sophisticated chemical agents, each one specific to one reaction, some of them accelerating their reactions by more than a trillionfold.[38] Our bodies harbor thousands of these catalysts. A good thing, too. Were one of them to falter, we would die, snuffed out like an early replicator without supplies.

But 3.8 billion years ago, protein catalysts had yet to be invented. Darwin's "warm little ponds" are poor sources of catalysts, which is one reason why many scientists became disenchanted with them.[39] Another is that two molecules have to meet before they can react. Because molecules are jostled erratically through water by heat's atomic vibrations, molecular encounters are chance events, and the chances are directly proportional to the number of molecules in a given volume of water: too few molecules, too few reactions. In other words, a metabolism can get going only if its molecules are concentrated. Dilute them in a bowl too large, and primordial life would end before it began. This is why chemists perform experiments in small test tubes and not in swimming pools. Washed out into the primordial ocean, newly created molecules would never be seen again.

Some think that tidal pools, a variant of Darwin's warm ponds, might solve this last problem. At low tide, water in such a pool evaporates through heat, thus concentrating chemicals. At high tide, new water flooding in can stir up the broth. But earth's violent youth casts doubt on this scenario as benign as a beach vacation. The moon orbited only a third as far away as it does today and tugged ferociously at the oceans, creating gigantic tides at least thirty times higher than today's. What is more, the moon positively whirled around the earth (which itself rotated twice as fast), circling it at least every five days, and would have created these extreme tides every few hours, leaving little time to concentrate life's ingredients.[40]

Evolutionary biology had been aching for better and smaller test tubes for decades, when an answer to its prayers arrived out of the blue— the deep blue. In 1977, the research submarine *Alvin* discovered an exotic menagerie on the Pacific seafloor near the Galápagos Islands, more than two thousand meters below the surface.[41] Red-plumed mouthless tube worms more than two meters long, snails with feet and shells armored with iron minerals, and eyeless shrimp thrived down there, cushioned by lawns of never-before-seen microbes that do double duty as food. But even more bizarre than this community is how it survives: Its raw materials come straight from Mother Earth herself, through searingly hot fissures in the earth's crust that overflow with nutrients, chemical energy, and the very catalysts that warm little ponds lack.

Cold ocean water percolates through these hydrothermal vents, sinking until it reaches the vicinity of giant magma chambers that heat it beyond the boiling point. From there it rises, like heated air in the atmosphere, until it is reunited with the cold waters above. On its journey through the deep-ocean volcanoes, this heated water leaches from the crust a thick broth of minerals, gases, and other nutrients. As it cools, they precipitate like snow from humid air. Unlike snow, however, they slowly aggregate into enormous chimneys whose height can exceed sixty meters. While growing, these chimneys continue to exhale a mix of hot water and small particles, and are thus aptly named *smokers*—black or white, depending on the chemicals in their breath.[42]

The hot water rising through hydrothermal vents may seem an obvious source of energy for life, but it is not the most important one—it's not the heat that makes a soup, but the ingredients. The vent fluids abound in energy-rich chemicals such as the hydrogen sulfide that gives rotten eggs their aroma. These volcanic chemicals would be pure poison to us, but they are fertile fuel to some microbes. Unlike plants, which *photosynthesize*—they extract energy from sunlight to build complex molecules from CO_2—these microbes *chemosynthesize*.[43] They build their own organic molecules from energy-rich inorganic molecules, as well as from the vent's abundant sources of carbon and other elements.[44] And theirs is not the only way to thrive around a vent. Although it is pitch-dark two thousand meters below the ocean surface—hardly any sunlight penetrates below two hundred meters—a heated vent also emits the faintest of glows, enough that some bacteria can scavenge its light for energy.[45] A vent's ways of provisioning life may be bizarre, but they are also highly effective, supporting oases with thousands of times more organisms than the surrounding seabed.[46]

Unlike the tepid soup of Darwin's pond, deep-sea hydrothermal vents are primordial pressure cookers. They are prevented from boiling over only through a kilometer-high water column that pressurizes them to some 200 atmospheres—equivalent to a mass of two million tons pushing down on every single square meter. Remarkably, even these extreme conditions do not deter life, as earth's current high-temperature champion testifies. The microbe called *Methanopyrus kandleri* can reproduce at temperatures above 122 degrees Celsius, which is higher than the temperature microbiologists use to sterilize their equipment.[47] (*M. kandleri* gets too toasty to reproduce only at 130 degrees, although it still survives there.)

Since Darwin's visit on the *Beagle,* the Galápagos Islands have been famous as an unusually fertile laboratory of evolution. This volcanic archipelago has already brought forth giant turtles, unique marine iguanas, and the playful Galápagos sea lion. So another seemingly unusual laboratory, the hydrothermal vents only 250 miles away, may seem a fitting if ho-hum companion. But the hydrothermal vents are not unusual. Thousands of them are fuming throughout the earth's oceans. They occur wherever magma rises from the earth's core and causes the seafloor to spread. And that is about everywhere along a giant chain of underwater volcanoes called the Midoceanic Ridge, a long wound reaching deep into the earth, bleeding liquid magma that constantly renews the planet's crust. A bit like the sutures on a tennis ball, this ridge circumnavigates the globe and is four times longer than the Rockies, the Andes, and the Himalayas combined, more than twice the circumference of the planet—all of it under water. Just as impressive as its length is the volume of water passing through the hydrothermal vents that litter this chain of volcanoes: More than 200 cubic kilometers every year, which means that all of the ocean's water circulates through one vent or another every 100,000 years.[48]

Hydrothermal vents have become favorite candidates for life's origin, but hardy and primitive life forms like this are not the most important reason. More important is that sources of energy and chemical elements are everywhere in their nutritious waters. Also, these vents are old, as old as the liquid oceans themselves. They have been exhaling nutrients since long before life began. Since then, all ocean water would have passed through them more than ten thousand times, enough to seed the oceans many times over.

Even better, hydrothermal vents solve several problems that plague warm little ponds. They provide exactly the needed test tubes, and in vast quantities. For the chimneys that rise when minerals precipitate from the hot rising water have a shape that is neither smooth nor simple. As the chimneys accrete from the precipitating vent fluid, they become suffused with numerous pores and channels, each one a minuscule test tube where microscopically small volumes of molecules can mingle and recombine without getting washed into the open ocean. Think of these chimneys as ever-growing laboratories stuffed with millions of tiny reaction chambers.[49]

As if that were not enough, these laboratories also come equipped with catalysts, not enzymes but minerals such as iron sulfide and zinc sulfide, some of them floating as particles in the vent fluids, others coating the surface of the reaction chambers.[50] And yet another benefit comes from the mixing of hot vent fluids and cold water. Both the reactions that build life's complex molecules and those that destroy them proceed faster when it's hot. The searing heat in a vent's core would render life's molecules unstable, whereas in the coldest areas around it life's reactions might proceed too slowly. But because waters of all temperature mix around a vent, a suitable temperature niche exists for any one of proto-life's chemical transformations.

Hydrothermal vents may well have been the laboratories that created the first metabolism. But even if we were sure of that—origin-of-life researchers don't agree on much—this knowledge by itself wouldn't specify *which* chemical reactions comprised the first innovation of life's history. The best candidates are the reactions found in the oldest parts of our own metabolism, those we share not only with other animals but also with plants and microbes, including the hardy ones around hydrothermal vents. Out of those possibilities, one candidate sticks out: a short cycle of chemical reactions called the citric acid cycle.

The citric acid cycle uses ten chemical reactions to transform one molecule of citric acid, the substance that gives lemons their sour taste, through several intermediates with uncommon names—pyruvate, oxaloacetate, acetate, and others—until it has completed one turn and manufactured another molecule of citric acid.[51]

A chemical cycle that creates two molecules from one sounds fishy, like the long-discredited perpetual motion machines of the nineteenth century. But this cycle does not violate any laws of physics. It cleaves the starting citrate molecule into two smaller molecules, from which its reactions build new molecules step by step, using as materials the carbon from carbon dioxide, and feeding on energy-rich nutrients.

Portions of the citric acid cycle appear in the planet's oldest known life forms, but its ancient heritage is not the only reason it is a prime candidate for the earliest metabolism.[52] The molecules that it creates are also ingredients for many other building blocks of life. Oxaloacetate provides atoms to build multiple amino acids and DNA nucleotides, pyruvate does the same for some sugars, acetate contributes to lipids—all-important components of cell membranes—and so on.[53] If you sought one metabolic core from which you could build what life needs, the citric acid cycle would be it.

What is more, the citric acid cycle is extremely versatile, for it can run in two directions.[54] In the first direction, described above, it operates a bit like an engine that performs the work of building new molecules, powered by a chemical battery of inorganic molecules. The kinds of bacteria that live in hydrothermal vents, bacteria that chemosynthesize for a living, use it in this way.[55] Run in the opposite direction, the cycle charges the chemical batteries that power life. Our bodies run it in this way to create chemical energy from the food we eat.

Even though the citric acid cycle's ancient heritage, source of building blocks, and versatility all advocate for its primacy, we are still waiting for an experiment like Miller's that would jump-start the cycle. Don't hold your breath. Such an experiment would be much harder than Miller's, because hydrothermal vents create such extreme conditions. Moreover, a chimney's reaction chambers have a complex shape and chemical coating that might have been the essential habitat for early life. You can't exactly order test tubes like this in the mail. But although we do not know yet how the entire cycle can emerge spontaneously, some experiments already point the way: With catalysts like iron sulfide and zinc sulfide, pyruvate, a key cycle molecule, has already been created spontaneously at high temperatures and pressures, and some of the cycle's reactions advance on their own in the laboratory.[56]

The citric acid cycle is attractive for one more reason: It makes more of itself. With each turn, it transforms a starting molecule into two, each of which spawns a new cycle and all its molecules, eventually creating four molecules, and so on. Chemists call this property *autocatalysis,* a fancy word for a defining feature of modern cells and primitive RNA replicators alike: They all make more of themselves.

The autocatalysis of the citric acid cycle differs from that of an elusive RNA replicase. Citric acid does not copy itself directly, nor do the cycle's other molecules. Instead they get copied indirectly through the entire network of reactions in the cycle. The hypothetical RNA replicase would be a self-replicating *molecule,* while the citric acid cycle is an autocatalytic *network* of chemical reactions. This isn't a shortcoming of the citric acid cycle, but another hint that a defining feature of life may not require RNA replicators and their genetic information: Life can exist before genes.[57]

We do not know—yet—whether the citric acid cycle is the grandfather of all metabolic activity. Nor do we know whether a metabolism of any sort came before RNA replicators. We do know, however, that the very first thing in the planet's history that deserves to be called alive needed an autocatalytic metabolism to still its hunger. Such a metabolism is more than a mere supply chain of parts, because each of its suppliers creates more suppliers, which can produce parts in ever-increasing numbers. And once both the factory and its supply chain are in place Darwinian evolution can kick in. It can preserve better factories, which demand improvements in the supplier, which permit better factories, and so on, in the unending cycle of evolution that lifts all boats.

It is perhaps more than a coincidence that hydrothermal vents can help close *this* cycle too. For they contain another curious catalyst called montmorillonite, named after the French town Montmorillon, where farmers use this clay mineral to retain water in drought-prone soils. Late in the twentieth century, the chemist Jim Ferris and others revealed another useful quality of montmorillonite, when they discovered that it can rally small RNA building blocks to assemble spontaneously into RNA strings more than fifty nucleotides long.[58]

Once metabolism and replication were in place, life was almost ready to crawl out of its cradle, but it still needed a travel bag. All of today's life uses the same kind of material to pack up its molecules, lipid molecules that are *amphiphilic,* from the Greek words for "both" and "love." An amphiphilic molecule "loves" both water and fat, because one of its ends likes to mingle with water, whereas the other avoids water—like oil that spreads in a thin film on a puddle. Observe lipid molecules in a solution and you are in for a surprise: They can form *vesicles,* minute hollow droplets enclosed by a tiny spherical membrane, in which the lipid molecules are arranged, as shown in figure 2.[59] How they could arrange themselves into complex highly ordered membranes without a guiding hand may seem mysterious but is not that hard to understand: This arrangement satisfies both parts of each molecule. The water-loving parts (solid circles

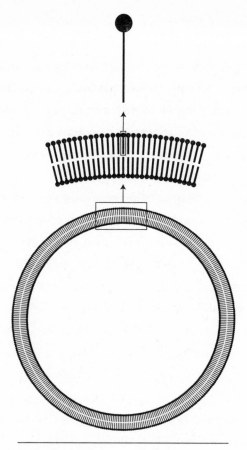

FIGURE 2. Biological membranes

in the figure) are close to water, whereas the water-avoiding parts (sticks in the figure) are away from it and close to each other. What is more, these membranes can grow spontaneously, incorporating new lipid molecules as you add them to the solution. And they grow autocatalytically: The larger they are, the faster they can grow.

To see where the building blocks of these membranes came from, we do not have to look far. The citric acid cycle produces one of their precursors, and they arise even in extraterrestrial rocks like the Murchison meteorite. Heat up powdered meteorite with water, and you will find molecules that self-assemble into vesicles.[60] What is more, montmorillonite, the same vent mineral that can string together RNA, accelerates membrane assembly. And hydrothermal vents can help in further ways, by concentrating membrane ingredients. This is what a team around Jack Szostak from Harvard University found when they re-created tiny reaction chambers like those of hydrothermal vents in their laboratory. They heated small amounts of lipids in tiny capillaries and saw that the lipids become concentrated at one end, until they start to form vesicles.[61] All by themselves.

It smacks of Van Helmont's spontaneous generation, all this complexity emerging from nothing but the right ingredients. But there is a crucial difference. Spontaneous generation—of mice, maggots, or microbes—requires the mysterious and perhaps supernatural vital force that Buchner's discovery of enzymes began to expose as an old wives' tale.[62] In contrast, the spontaneous creation of membranes and molecules—self-organization in modern language—requires only mundane physics and chemistry. Assembling membranes requires nothing but the attraction of similar molecules. Like the self-aggregation of volcanic particles into towering underwater buildings, or the spinning of RNA strings by clay minerals, the self-organization of both membranes and molecules is explained by well-known laws of nature.

Self-organization permeates the universe so completely that most of us don't even notice it. Much older than life and natural selection, self-organization is how stars and solar systems form, how the earth accreted, how it acquired a moon, oceans, and an atmosphere, and how the continents started to shift. Self-organization creates the microscopic symmetry of a snowflake and the raging clouds of a hurricane, the shifting shapes of sand dunes and the timeless beauty of a crystal. We shouldn't be surprised to find self-organization in life's precursors, because it is everywhere else too.

Life's self-organizing membranes can solve another one of early life's puzzles: the mechanism by which the first cells divided. Modern cells use very sophisticated machinery—dozens of proteins—to constrict and divide cells, and to make sure that each daughter cell receives a copy of the mother's DNA. But they could do the trick in simpler ways, as Szostak's team found in 2009. The researchers observed how rapidly growing membrane droplets change their shape when they divide, and transform into threadlike hollow tubes. Unstable tubes, I should say. Agitate them a bit and they fragment spontaneously into smaller droplets. Even better, when the researchers placed RNA molecules inside these tubes, they were partitioned among the droplets. Lifeless membrane droplets can divide like living cells—an innovation without an innovator, emerging from a simple property of the system's chemistry. All by itself.

Although we have come a long way from the first musings about a primordial soup, there are some problems that still defy solution. One of them is the last obstacle on the path from a self-dividing membrane droplet to a primitive cell. If the RNA inside this cell replicated faster than the cell grew, it would divide until the vesicle was ready to burst. But if the cell outpaced the RNA in growing, the RNA inside would become increasingly dilute, and many droplets would spawn empty-shelled offspring. To succeed, life needed to balance, to *regulate* replication and growth with precision, such that RNA replicated no faster than its container grew. How it learned to do that remains a mystery that twentieth-century science has left for another generation.

Fast-forward from the first wheel to the Ferrari. Although some of life's features did not change in the thirty million centuries since it began—molecules, regulation, and metabolism are still wellsprings of innovation, as we shall see in later chapters—evolution transformed just about everything else about it. Early RNA replicators have been replaced by complex protein machines. Life has learned to regulate not just RNA and lipids but thousands of molecules. And innumerable innovations have turned the metabolism of a modern cell—the Ferrari's engine—into a miracle of chemical technology.

Imagine driving home in this Ferrari from an evening picnic and running out of gas on a lonely stretch of highway in the middle of the night. No gas station is in sight, nor is anybody you could hitch a ride with. But no matter. You open your trunk, where a cooler contains left-over food and drinks. You pour a bottle of orange juice into the tank, and after it a quart of milk, and then a glass of wine. That will be enough to tide you over to the next gas station. And on you drive.

Modern metabolic engines are just like that. They can run on many different fuels. And more than that, they can also use each fuel as raw material to manufacture the smallest molecular parts of their body, parts the body needs to grow, to reproduce, and to heal. It's as if a car could use the stuff in its gas tank not only to operate the engine but also to patch a leaky tire or mend a broken windshield.

The molecular parts in question comprise a few core molecules, some sixty *biomass building blocks* from which our bodies are con-structed and repaired.[63] The most important are the four DNA building blocks of our genome, the nucleotides composed of a sugar, a phosphate group, and one of the four nitrogen-containing bases adenine (A), cyto-sine (C), guanine (G), or thymine (T). Next are the four building blocks of the RNA into which this DNA is transcribed, and that still controls much of life. They—A, C, G, and U, for uracil—differ only in a single oxygen atom from DNA building blocks, but this single atom makes a huge chemical difference. It makes RNA the better catalyst, and DNA the better—because more stable—information repository. Then there are the twenty amino acid building blocks of the amino acid strings translated from RNA, some of them familiar, like the tryptophan blamed for post–turkey dinner drowsiness, or the glutamic acid of the flavor enhancer monosodium glutamate (MSG). Together with the lipids in membranous bags, some energy storage molecules for hard times, and molecules that help enzymes do their job, these comprise the sixty different kinds of bricks from which cells build themselves.

The tasks of metabolism—procuring energy and making stuff—have not changed in the last 3.8 billion years. And neither has its basic nature, a network of chemical reactions like the one where the white table sugar sucrose reacts with water and splits into the two more digestible molecules glucose and fructose. What has changed is the *number* of those reactions. Our earliest ancestors got by on a handful of reactions, but modern metabolism, like modern life in general, is much more complicated.

Modern metabolism is an interlaced, highly connected network of chemical reactions, the product of four billion years of innovation. If you were to chart it out, it would resemble a map of every street in the United States, from the shortest residential cul-de-sac to the complete interstate highway system. At its core is the ancient citric acid cycle—as central as Pennsylvania Avenue, which connects the White House with the U.S. Capitol. Figure 3 shows a tiny sliver of such a network, whose lines connect different molecules (shapes) that react with one another. Think of it as the road map of a village. The four molecules involved in cleaving white table sugar are written out and encircled within an ellipse. But don't

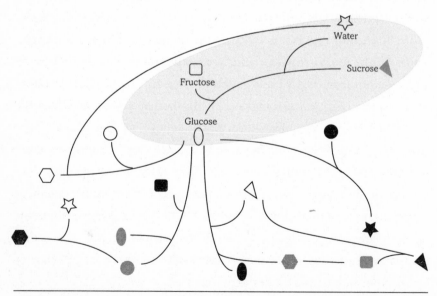

FIGURE 3. A tiny sliver of a metabolic network

take this visual crutch for the real thing. Fructose can participate in thirty-seven reactions rather than the single one shown, and many more molecules and reactions are needed to run a modern metabolism.

To find out how many required more than a century of research. During this time, thousands of biologists built a tower of knowledge about metabolic reactions by studying the human gut bacterium *Escherichia coli*. Its construction took about as long as a medieval cathedral, but the vista from the top is spectacular. We now know how *E. coli*'s metabolism—more than a thousand small molecules that rearrange themselves in thirteen hundred metabolic reactions—is wired.[64] And we know that in the metabolism department *E. coli* and many other microbes beat us hands down. For example, of the twenty amino acids in our proteins, our bodies can only manufacture twelve. The other eight we have to get from food. In addition, we need thirteen vitamins to live, but can synthesize only two of them, vitamins D and B_7 (biotin).[65] *E. coli* can cook all of them up from scratch.

Part of the reason why *E. coli*'s metabolism is so complex lies in the sixty-odd biomass building blocks. Manufacturing each of them requires multiple reactions and intermediate molecules. Another part is that *E. coli* is a phenomenal survivor, thriving not only on the rich nutrient broth of our guts but also in an austere nutrient desert where only seven small molecules supply chemical elements and energy. This minimal environment is so spartan that one molecule like glucose does double duty as a source of both energy and the chemical element carbon. From these few ingredients, *E. coli* can manufacture everything it needs, all sixty-odd biomass building blocks, and from them, the entire cell.

But that's not all. You can remove glucose from a minimal chemical environment and replace it with another source of carbon and energy, such as glycerol. *E. coli* can still build its body from the carbon and energy in this molecule. Replace glycerol with the acetic acid in vinegar and, again, *E. coli* can build its body. All in all, *E. coli* can use more than eighty different molecules as its *only* source of energy and as its only supplier of every single one of the billions of carbon atoms in its cells. It is similarly flexible about other elements, such as nitrogen and phosphorus. *E. coli* is like a self-building, self-multiplying, self-healing race car that can run on kerosene, Coca-Cola, or nail polish remover.

Simple chemical environments are useful for studying microbes in the laboratory, but they are rare in the wild. An environment like the soil or the human gut contains dozens of ever-changing fuel molecules. To harvest energy and to extract building materials from these molecules require a distinct sequence of chemical reactions for each of them. And to make a good living, a microbe must be able to exploit all of them.

Suddenly, a thousand reactions don't sound like a lot.

Another difference between today's life and its shadowy ancestors lies in the catalysts, those molecules that accelerate chemical reactions. If your gut did not contain the right catalyst—an enzyme known as *sucrase*—the sucrose in a drink of sugared water would take years or decades to split into glucose and fructose.[66] You could drink gallons of sugared water every day, and starve to death.

Reactions like this are no longer accelerated by the simple metal-containing mineral catalysts of early life. Modern catalysts speed up some reactions a trillionfold, allowing molecules to react as soon as they meet. Each of these molecular machines—and there are several thousand of them—is a specific string of amino acids.[67] The enzyme *sucrase*, for example, is a gigantic molecule with 1,827 amino acids, each of them with at least a dozen atoms, adding up to twenty thousand atoms per sucrase molecule.[68] The table sugar sucrose with forty-five atoms is minuscule by comparison—like a pea compared to a football—which explains why enzymes are called *macro*molecules, as opposed to the small molecules they help react and the biomass building blocks they help construct.[69] Sucrase may seem large, but it is not even unusual. Many enzymes are much larger.

While the sucrase string is manufactured, it curls and twists in three dimensions, like a ball of wool, but with important differences: Every ball of wool is unique, but every sucrase molecule is the same. As sucrase is manufactured, it folds in space in a precisely stereotypic manner. What is more, folded sucrase is constantly wiggling, jiggling, and vibrating to perform its catalytic duty. Think of sucrase as a self-assembling nanomachine whose movement is so fast it would be a mere blur, taking molecules in, cleaving them, and spitting out their products at lightning speed.

Every cell contains thousands of such nanomachines, each of them dedicated to a different chemical reaction. And all their complex activities take place in a tiny space where the molecular building blocks of life are packed more tightly than a Tokyo subway at rush hour. Amazing.

We do not yet know how life evolved all this complexity from its simple origins, and we may never know for sure. The oldest single-celled fossils are as complex as modern cells, and their ancestors are shrouded in darkness. This should come as no surprise. The eons have ground away most ancient rocks, and even if the churning continents had not liquefied their remnants, early life was a fragile bag of molecules. It was nothing like the sturdy mats of blue-green algae—more correctly called cyanobacteria— that left behind 3.5-billion-year-old calcium imprints known as stromatolites, and even less like the big-boned dinosaurs who lived a relatively recent hundred million years ago.

We do know, however, that we all come from a single common ancestor. This is not the same as saying that life originated only once. Given the powers of self-organization, I would not be surprised if life arose many times, in hydrothermal vents, in warm ponds, or who knows where else. Among a multitude of faint lights that flickered on and off throughout the earliest history of the planet, some held steady, while others shone more and more brightly. But only one of them became bright enough to spawn all of today's life. This is not a matter of opinion. It has to be true, for a single reason: standards. More accurately, *universal standards*.

The computer scientist Andrew Tanenbaum once quipped, "The nice thing about standards is that you have so many to choose from."[70] I know what he was talking about. Whenever a remote control, a clock, or some other gadget stops working in my home, I rummage through a cabinet in my living room that contains a zoo of batteries large and small—but usually not the right one. Life would be easier if it offered only one kind of battery. Or one kind of coffee filter, data storage medium, or computer operating system. Even old technologies suffer from this problem: After more than a century of public electric power, fourteen incompatible outlet standards exist around the world, a curse for millions of international travelers who arrive in foreign countries every day accompanied by laptops, hair dryers, electric razors—and the wrong outlet adaptors.

Nature is different. It has standardized energy storage. Among the many forms that energy can take, such as mechanical (a wrecking ball smashing into a house), electrical (the current of electrons powering a computer), or chemical (the bonds that tie atoms together in a molecule), chemical energy is life's favorite. All organisms on the planet, from single-celled bacteria to the blue whale, use a standard means to store energy, the molecule adenosine triphosphate (ATP). When its energy-rich chemical bonds rupture, energy is transferred to other molecules, and the less energy-rich molecule adenosine diphosphate (ADP) is created. To regenerate the energy-rich ATP, specialized enzymes can transfer energy to ADP from fuel molecules.

Not all of ATP's chemical energy ultimately ends up in other molecules. Bacteria use ATP to power the tiny whirring flagellae that propel them through water. Fireflies use ATP to illuminate their bodies when they hope to attract mates. Some eels transform ATP into electrical energy that dispatches prey with powerful electric shocks. But regardless of its final form—mechanical, light, electric—the energy in living things ultimately comes from the chemical battery of ATP.

When a cell uses a chemical fuel like glucose to manufacture one of the cell's biomass building blocks, it first converts the chemical energy from glucose into the chemical energy of ATP. It then uses ATP's chemical energy to build, step by step, the chemical bonds of the building block. In this way, the energy stored in the fuel eventually ends up in the bonds of the building block. ATP is a crucial middleman in this energy transfer.

Living things have adopted ATP as the universal energy storage standard—no rummaging for batteries or paying a premium for an airport power adapter.[71] Every organism living today can trace its descent from the inventor of life's most successful power storage innovation. And power storage is not life's only standard. We have already encountered the ancient heart of metabolism, the citric acid cycle, and the universal membrane molecules with their love-hate relationship to water.[72] And let's not forget DNA, RNA, and the genetic code that translates triplets of DNA letters into amino acids—a code understood by all organisms.[73]

ATP and the citric acid cycle aren't universal standards in the same way that the speed of light is a universal speed limit. They aren't the *only* way to build life. We know alternatives to our genetic code, to ATP as an energy carrier, and even to DNA as an information repository.[74] Life's standards are the historical legacy of a single ancestor. The marathon that started at life's origins may have begun with many hopeful participants, but whether through natural selection or dumb luck, only one crossed the finish line to leave its descendants today. This is a bit depressing, if you extrapolate from the present to your chances of leaving descendants in the distant future. But it also contains a hopeful message, at least for frequent travelers: Wait another four billion years, and you may not need an outlet adapter.

By the time you read these lines, the puzzle of life's origin may be complete. We may know whether life began in a warm pond, in a hydrothermal vent, in a freezing ocean, or in outer space. Or we may have to wait another century. But more important for understanding innovability than reconstructing the one true scenario are two general lessons that all scenarios have in common.

The first is that life needed to innovate even before it became life—by creating the first autocatalytic metabolisms and the earliest replicators.

The other is that life's symphony of innovation has three major themes. First, innovations created new combinations of chemical reactions, such as those that form life's building blocks and that built the first replicators. Second, innovation required molecules that could help other molecules react. Third, innovation created new regulation, the key to coordinate complex life. These three themes resounded louder and louder in the biosphere as life became more and more complex and innovability increased. Primitive metabolism has grown into a giant network in which chemical reactions are combined and recombined to permit life's expansion into every conceivable habitat. Sophisticated protein molecules have pushed aside simple inorganic catalysts, and have given rise to innovations as different as light-detecting opsins and armor-providing keratins. And regulation, a seemingly mundane process, has become an innovation industry all by itself, bringing forth multicellular organisms with limbs, a heart, and a brain.

From the origin of life to today, innovations have been transforming metabolism, proteins, and regulation. And although the three seem very different, a curious but powerful kind of self-organization stands behind their ability to innovate.

CHAPTER THREE

The Universal Library

I magine standing in a room crammed with books from floor to ceiling. The bookshelves barely leave space for the door you see on each of the four walls. You start leafing through the books and realize that they all have the same number of pages. Each page contains the same number of lines. And each line has the same number of characters. But—this is strange—the books are full of gibberish. Each line of each page of each book contains mostly arbitrary strings of letters—"hsjaksjs . . . ," "zvaldsoeg . . . ," and so on—occasionally separated by spaces and punctuation. Only rarely do you find a meaningful English word—"cat," "teapot," "bicycle"—islands in a vast sea of more gibberish.

After a while you tire of these books, which do not make sense. You step through one of the doors and find yourself in another room just like the first one. It is equally packed with bookshelves that crowd in on four doors. And its books make no more sense than those in the first room.

Another door leads you to yet another identical room, and from there you begin to wander through room after room after room, and realize that you are in an endless maze of rooms, identical except for the books that inhabit them. These books form a library that is as gigantic as it is bizarre.[1] As you wander through this library, you encounter fellow travelers who help you grasp the enormity of this place.

The rooms form a universal library, home to all conceivable books.

That is, its books contain all possible strings of characters—twenty-six letters and a few punctuation marks. Most of the strings are the nonsense you already read. But occasionally a book will contain a meaningful word, sentence, or paragraph. More than that, somewhere in this library dwell books that contain no gibberish whatsoever. Because the library contains all possible books, it also contains each meaningful book ever written. All possible novels, short stories, poetry collections, biographies (of people real or imagined), philosophical treatises, religious books, books of science and mathematics, all conceivable books written not only in English but in all languages, books that reveal everything that is true, but also spin terrible lies, books that talk about other books, about the library itself and where it came from, books, some true, others false, about your life's story, how it began and how it will end, and the book you are reading right now. All of them are contained in this library—a library enormous almost beyond imagining.

To get of an idea how large this library is, let us say that every book in it contains 500,000 characters. (That's not very long—in the same ballpark as the book you are reading right now.) Excluding punctuation marks, there are 26 possibilities (A through Z) for each of these 500,000 characters. That is, there are 26 possibilities for the first character, 26 for the second, 26 for the third, and so forth. To estimate the number of books, we thus need to multiply 26 by itself 500,000 times. Mathematicians would write this number as 26 raised to the power of 500,000, or 26^{500000}. This is a very large number, amounting to a 1 with more than 700,000 zeroes behind it, more zeroes than this book has letters. And far greater than the number of hydrogen atoms in the universe. It is a hyper-astronomical number.

The deepest secrets of nature's creativity reside in libraries just like this: all-encompassing and hyperastronomically large. Only instead of being written in human language, the texts in these libraries are written in the genetic alphabet of DNA and the molecular functions that DNA encodes.

Human books can capture entire universes—everything that human language can utter—but they have nothing on the chemical language of what may be life's oldest library of creation, the one devoted to metabolism. Every one of the trillions of living things on earth can be *described* by human prose or poetry. But *creating* any one of them requires the chemical language of metabolism, the chemical reactions that create the building blocks of life and thus ultimately all living matter. The library's chemical language can express life itself—all of it.

To date, we have discovered more than five thousand different chemical reactions that some organism, somewhere on our planet, uses to produce the building blocks of life I mentioned in chapter 2, the nucleotides that make up DNA and RNA, and the amino acids from which proteins are constructed. The reactions that occur in *E. coli*—more than a thousand—are among them, as are all known chemical reactions that take place in any bacterium, fungus, plant, or animal—including humans. When your body extracts energy from sugar or any other food, it uses such reactions. It also uses them when healing the few hundred skin cells covering a scraped knee, and when replenishing the millions of red blood cells that die every day.

No organism can catalyze all five-thousand-odd known reactions, but every organism can catalyze *some,* and the reactions it *can* catalyze make up its metabolism. For multiple organisms we know these reactions, thanks to twentieth-century biochemistry and to the technological revolutions of the early twenty-first century. They gave us access to a mountain of metabolic information on more than two thousand different organisms, stored in giant online repositories, such as the Kyoto

	Reaction Universe	Genotype
	Sucrose + Water → Glucose + Fructose	1
	A + B → C + D	0
	.	.
	.	.
	.	.
List of 5,000 reactions	D + E → F + G	0
	H + I → J + K	1
	.	.
	.	.
	.	.
	I + K → L + M	1
	N + P → O + Q	0

FIGURE 4. A metabolic genotype

Encyclopedia of Genes and Genomes, or the BioCyc database, and accessible in split seconds from any computer with an Internet connection.[2]

Figure 4 shows how we can organize this information. The left side of the figure stands for a list of five thousand reactions—written as chemical equations. To avoid clutter I wrote out the molecules in only one of them—the sucrose-splitting reaction—but simplified all others to a single letter. Let's consider one organism, such as *E. coli* or a human, and mark a "1" next to a reaction if our organism can catalyze this reaction—it has a gene making an enzyme for it. Otherwise we'll mark a "0." The result is a long list of ones and zeroes like that in the figure, a compact way to specify a metabolism.

Bacteria such as *E. coli* can make all twenty amino acids in proteins, whereas metabolic cripples like us humans can make only twelve of them. We lack the necessary enzymes and reactions for the remaining eight. The figure's shorthand way of describing a metabolism is ideal for expressing differences like this: Because we lack some reactions, our list of reactions contains some zeroes where that of *E. coli* contains ones.

A list like this is also an extremely compact way to write an organism's *metabolic genotype*—the part of the genome encoding its metabolism—because an organism's list of reactions is ultimately encoded in its DNA. You can also think of the list as a text written in an alphabet with only two letters, and without spaces or punctuation marks, like this: "1001 . . . 0110 . . . 0010." The first letter in such a text might correspond to the sucrose-splitting reaction, which is present ("1") in this example, whereas the second reaction might be one of those needed to synthesize an essential amino acid—it is absent ("0") in this example text but could be present ("1") in another organism's genotype—and so on.

It is a text in a library vast beyond imagination, the library of all possible metabolisms.

The number of texts in that library can be calculated with the same arithmetic that computed the size of the universal library of books. Because each reaction in the known universe of reactions can be either present or absent in a metabolism, there are two possibilities (present or absent) for the first reaction, two for the second reaction, and so on, for each reaction in the universe. To calculate the total number of texts, we multiply the number 2 by itself as many times as there are reactions in our universe. For a universe of 5,000 reactions, there are 2^{5000} possible metabolisms, 2^{5000} texts written in the alphabet of zero and one, each of them standing for a different metabolism. This number is greater than 10^{1500}, or a 1 with 1,500 trailing zeroes. While not quite as large as the number of texts in the universal library of human books, it is still much larger than the number of hydrogen atoms in the universe. The metabolic library is also hyperastronomical.

And just as the universal library contains all meaningful books, the library of metabolisms contains all "meaningful" metabolisms—those that allow an organism to survive—and many more, because not all metabolisms are meaningful, just as not all books are. Some metabolisms cannot procure energy, or they fail to manufacture important molecules. These are like books where some chapters, paragraphs, or sentences are coherent but the book as a whole does not make sense. And many other metabolic texts are gibberish. These are metabolisms with disjointed reaction sequences dead-ending on molecules useless to life, the equivalent of books containing only meaningless character strings.

If you wandered through the universal library of books long enough, you would find books that surprise you. They contain novel thoughts, ideas, and inventions. The genotypic texts in the universal metabolic library are no different. They can encode metabolisms with never-before-seen chemical abilities, novel phenotypes that manufacture new molecules or use new fuels. In short, innovations.

Because metabolism is as old as life itself, evolving life has explored this library ever since it originated. A billion years ago, nature had already discovered unimaginably many metabolic phenotypes, enough of them that it might have stopped finding innovative metabolic texts long since. But far from resting on its early laurels, evolution is still discovering such texts, much faster than we can decipher them, in billions and trillions of organisms alive today. Some of these texts appeared less than a hundred years ago—a mere moment in evolutionary time.

Consider pentachlorophenol, a nasty molecule that humans first produced in the 1930s. It is used in antifouling paint to coat ships' hulls, and also as an insecticide, fungicide, and disinfectant—in short, to kill life. Pentachlorophenol also damages our kidneys, blood, and nervous system, and it causes cancer. But despite its noxious nature, life has found ways not only to tolerate pentachlorophenol but to thrive on it. The aptly named bacterium *Sphingobium chlorophenolicum* can extract both energy and carbon from it, using pentachlorophenol as its only food source. To do so, its genome encodes four enzyme-catalyzed reactions that convert pentachlorophenol into molecules that are as digestible as glucose—the equivalent of transforming a chemical weapon into a chocolate bar.[3]

The combination of these reactions is unique to *S. chlorophenolicum,* but the reactions themselves are not. Each of them occurs in hundreds if not thousands of other organisms. Two of them help recycle superfluous amino acids in some bacteria, whereas the other two disarm toxic molecules produced by some fungi and insects—molecules that happen to resemble pentachlorophenol.[4] Like a garage mechanic building a sprinkler system out of an alarm clock, a bicycle pump, and some PVC pipe, evolution has created in *S. chlorophenolicum* a new arrangement of chemical reactions catalyzed by enzymes that individually exist in other organisms. In other words, metabolic innovation is combinatorial.

Innovations that allow organisms to feed on highly toxic, man-made molecules are not rare. The bacterium *Burkholderia xenovorans* happily feeds on the now outlawed polychlorinated biphenyls, which were widely used in making plastics and in the electrical industry.[5] Other bacteria readily digest chlorobenzene, a toxic organic solvent used in chemical laboratories.[6] And even more striking are the bacteria that feed on the very antibiotics designed to kill them.[7] Some of these antibiotics are man-made, so bacteria did not encounter them until recently.

Just as nature can convert poisons into food, it also came up with ingenious ways of managing its waste. Ammonia (NH_3), for example, isn't just the gas in household cleaners with the sharp, unpleasant odor that makes your eyes burn, but a highly toxic waste product of animal metabolism. Because ammonia dissolves in water, fish can just excrete it into the water surrounding them and forget about it—the fish equivalent of peeing in the swimming pool. But when animals first conquered land more than three hundred million years ago, they did not have this luxury. They needed to prevent toxic ammonia gas from poisoning their blood.

The solution lay in a metabolic text that contains the instructions for converting ammonia into the less toxic molecule urea, which we secrete to this day in our urine. This metabolic innovation involves five common chemical reactions, each one independently useful to organisms long before the need to detoxify ammonia appeared.[8]

When exactly this innovation appeared is unknown, but clues are easy to find. Even though modern *teleosts*—bony fish—have no need to detoxify ammonia, *their* ancestors already harbored a chemical blueprint for making urea, still seen in cartilaginous fish like sharks and rays that swam through the oceans long before modern fish appeared. However, the title character of *Jaws* uses urea for a different purpose than the humans hunting it—for nitrogen storage, buoyancy, or as a counterweight to the salt in seawater. (You might think that the DNA of bony fish should contain some remnant of this innovation, if it had already originated in their distant ancestors. And that's indeed the case: The text for the urea cycle still exists in bony fish, even though they rarely express its chemical meaning. They are a bit like adults who learned a language while infants and can still recognize some of its words.)[9]

Detoxifying your waste is good, but recycling it is even better, and nature excels at that too. The nitrogen waste of animals—ammonia or urea—fertilizes plants. The very oxygen we breathe is a waste product of photosynthesis.[10] And every gram of feces is teeming with billions of bacteria feeding on the molecules in it: One man's waste is a bacterium's treasure. Each one of these bacteria harbors a metabolic text, ancient or recent, to break down molecules, extract energy and chemical elements, and build new life from them.

Innovative metabolic texts are just as ubiquitous in extreme environments—extremely hot, extremely cold, excessively dry, highly caustic, exceedingly radioactive, super-salty, and so on—as in temperate ones. Bacteria in particular can thrive in boiling water and in ice, in highly corrosive sulfuric acid and in crushing oceanic depths. To survive, they had to innovate, and many of their innovations—you guessed it—are metabolic.

Without these innovations, extreme environments would kill bacteria just as easily as they kill us. Too much salt, for example, kills cells, because it forces water out through osmosis and prevents enzymes from doing their job—they need water as a lubricant. To clog this drain, metabolism produces molecules with exotic names such as ectoine and glycine betaine that cannot leave a cell as easily as water does, and that can stand in for water molecules lost through osmosis. They keep proteins lubricated. To make these molecules, cells need only a few extra enzyme-catalyzed chemical reactions that start with common molecules like the amino acid aspartate. Add these reactions to a metabolism, and you have a leg up in the most hostile environment. Halophilic bacteria—the name comes from the Greek for "salt-loving"—can survive salt concentrations of 30 percent, ten times higher than the seawater that kills us when we drink it. They can even live around and inside salt crystals.[11]

Extreme environments are no picnic, but life can be even harder if you face predators and parasites, and especially if escaping them is not an option. Any ordinary plant would be an immovable feast for many organisms, from insects and worms tunneling through the soil to slugs and other herbivores aboveground. Because plants can't so much as twitch in their own defense, they develop chemical weapons, molecules so toxic that animals avoid them. Plants are not alone in using chemical warfare, but they are especially adept at it, perhaps because they are, literally, rooted to one spot.

These defensive molecules are metabolic innovations, because they require new combinations of chemical reactions to synthesize them. One of them is the nicotine produced by tobacco plants that some of us blissfully inhale through cigarettes, even though it is so toxic that some farmers use it as an insecticide. But plants had the idea first, as a group of German scientists recently showed. When they artificially lowered the amount of nicotine that tobacco plants produce, insects developed a voracious appetite for the plants. They attacked the plants more often, ate more leaves, and grew faster. The plants, in turn, lost three times more leaves than normal plants to their attackers.[12]

Nicotine is only the best known of more than three thousand similar alkaloids—a catchall term for organic molecules built around nitrogen atoms, including caffeine and morphine—that plants use in chemical defense. And although they are numerous, alkaloids are only one among several kinds of chemical warfare molecules. Others include the astringent tannins that make your mouth feel dry and shriveled when you eat unripe fruit.[13] Tannins bind very tightly to plant proteins and prevent our gut from digesting these proteins, which discourages us from feeding on them in the first place. Cyanogenic glycosides are especially nasty chemical defense molecules that are produced in cassava or manioc, the important African and South American food plant.[14] Unless you remove these glycosides by cooking or soaking, they release hydrogen cyanide, the active ingredient of the Zyklon B pesticide piped into the "showers" at Nazi extermination camps like Auschwitz-Birkenau. If you ever thought of nature as that idyllic place, the next best thing to the Garden of Eden, a tutorial on chemical warfare in plants will quickly dispel that myth.

Biochemical warfare molecules like these are metabolic innovations, add-ons to an existing metabolism manufactured by new sequences of chemical reactions that start from common biomass molecules and transform them into potent poisons. Each one requires specific passages of text in a metabolic genotype.

Some of nature's ways to find new metabolic texts are familiar, because they dominate in large multicellular animals like us. They include the changes accompanying sexual reproduction, which shuffles chromosomes like decks of cards, so that each of our children starts with a new deal. Then there are the spontaneous mutations in a DNA's letter sequence, arising through chance events such as when photons of ultraviolet radiation smash into the genome, or through highly reactive oxygen radicals that are by-products of chemical reactions and burst the chemical bonds of nearby DNA.

Neither way to explore the metabolic library is very effective. Since the shuffling of sexual reproduction occurs between highly similar genomes—two human genomes share 99.9 percent of their DNA letter sequence—it is not the most effective way to create new metabolisms.[15] It is like trying to write a new play by changing thirty words in *Hamlet*. And while mutations can create new proteins, including new enzyme catalysts, they are rare, which means the process is rather slow.

And there is one more reason why metabolic innovation is not swift in large, multicellular animals. A new way of using energy or building organic structures can make its value known only at the speed that it spreads throughout a population, and animals that produce a new generation every few decades—or even every few months—can't innovate any more rapidly than that.

All this doesn't mean that animals like us are completely impoverished when it comes to metabolic innovations. Our bodies, for example, can disarm drugs—like the widely used aspirin, known to chemists as acetylsalicylic acid—through a metabolic process called glucuronidation that renders them less toxic and excretable in urine. Cats and some other carnivores like hyenas lack this enzyme.[16] (Consult a vet before medicating your pet hyena with aspirin.) You may ask why our bodies have this enzyme, which evolution created long before the company Bayer first marketed aspirin in the 1890s. The clue lies in aspirin's name itself, which comes from an old name for a plant called meadowsweet, *Spiraea ulmaria*. This and many other plants have been used since antiquity for pain relief. What is more, plants containing salicylic acid were part of our ancestors' diets, such that our omnivoric bodies—unlike those of carnivores like hyenas—needed a way to detoxify it.

Within the multicellular world, humans are far from the pinnacle of metabolic creation, however, because many animals beat us in other aspects of metabolism. Humans cannot produce vitamin C, and must therefore drink it with a morning glass of orange juice, whereas dogs can make their own. And although we can extract calories from the seeds of grasses like wheat and maize, cows are better at digesting the cellulose in their stalks. To be fair, however—credit where it is due—that miracle of metabolism isn't really a bovine innovation, but a microbial one: It is the bacteria in the four-stomached cows that convert gigantic cellulose molecules into easily digested glucose.

Which is a hint that the real geniuses of innovation are the smallest organisms on the planet: bacteria.

This isn't just because bacteria produce new generations in minutes rather than years, and so can improve their genetic toolkits much faster than we. The innovation advantage of bacteria goes much deeper than that. To grasp how big it is, imagine a teenage boy trying to make his high school basketball team, even though he's only a shade over five feet tall. Hard work and exercise can only take him so far. He just doesn't have the right genes—not like his best friend, who can practically touch the rim on his tiptoes.

A bacterium wanting the bacterial equivalent of a forty-inch vertical leap isn't limited by the genes bequeathed it by previous generations. If, in some science-fiction movie, perhaps, the two basketball-playing friends had the same innovation equipment as bacteria, the process would look something like this. Our two characters are dining in their favorite restaurant when a slender hollow tube begins to grow out of the taller boy's body, blindly groping toward the shorter. As soon as it connects, this tube injects a random fragment of the taller guy's DNA text into the other body, and if this DNA contains the right genes, the high school basketball team gets a new power forward.

This is an example of horizontal gene transfer, a phenomenon tragically unavailable to disadvantaged humans but rampant in microbes. Sometimes when two bacteria are in proximity, one of them will extrude a slender stalklike hollow tube in the direction of the other. When the tube docks, it shrinks in length, draws the two cells closer, and through the tunnel thus created, one cell transfers DNA to its neighbor.

This transfer resembles sex as we know it, because a penislike tube transfers genetic material from one organism to another. But bacterial and human sex are quite different. Their sex, unlike ours, does not serve reproduction. And it doesn't even shuffle a whole genome, but usually just transfers a few genes.

Bacteria can acquire new genes in other ways too. Some absorb DNA from other cells after they die, rupture, and spill their molecular innards. As boneheaded as a person who would rather eat books—surely a great source of fiber—than read them, they use some of this DNA as food. Occasionally, though, the eaten DNA becomes hitched to their genome and helps make new proteins.[17]

Gene transfer can also take advantage of viruses, those tiny lifeless particles whose DNA can enslave cells many times their size.[18] While viruses reprogram a cell into a helpless factory that clones its viral masters, small pieces of the cell's DNA can fuse with the viral genome. These pieces piggyback on newly minted viruses that leave the cell, and get injected along with the viral genome into the next hapless victim. In this scenario, one of our basketball-playing friends might simply have to sneeze on the other, whereupon his talent could be transferred to his teammate's newly improved genome.

If all this horizontal gene transfer went on unchecked, the size of a genome would constantly increase over time and become grotesquely bloated. But excessively long DNA strings break more easily, and copying them wastes energy and materials—a mortal sin that nature won't tolerate.[19] Fortunately, such bloating does not occur, because gene transfer is balanced by gene deletions. These are the by-products of errors that happen when cells cut or splice DNA molecules as they repair and copy their DNA. Unlike DNA mutations that alter one letter at a time, deletions can strike out thousands of letters and many genes. As long as a deletion affects no essential genes, the cell can live with it. Such survivable deletions occur all the time. They ensure that only useful genes stay around in the long run, and they keep genomes lean.

In another difference from sex as we know it, gene transfer occurs not just between similar organisms but also between baker's yeast and fruit flies, between microbes and plants, and especially among bacteria, which can be as different from one another as humans are from oak trees.[20] This is why gene transfer is so powerful, and the most important reason why bacteria are masters of metabolic innovation. Very different organisms harbor very different metabolic texts, and gene transfer can edit one text with borrowed passages that are very different yet meaningful within another text, the microbial equivalent of a musical mash-up that combines a Baroque instrumental track with a pop vocal. Only some edits will improve a text, because the recipient cannot pick and choose which new genes it gets—they are a random subset of the donor's genome. But because gene transfer is incredibly frequent, the odds for innovation aren't bad: Even though many edits lack luster, the shelves of life's universal library contain a virtually infinite number of masterpieces waiting to be found.

An example of nature's editorial prowess is our friend *E. coli* and its multiple varieties, *E. coli* strains that were long thought to be like closely related ethnic groups.[21] At the beginning of the twenty-first century, biologists first deciphered the genomes of many such strains, expecting them to be very similar. False. Two *E. coli* strains can differ in more than one million letters, or one-quarter of their DNA, such that one strain can harbor a thousand genes that the other strain lacks.[22] Every million years—a blip of evolutionary time, 20 percent of the time since humans diverged from chimpanzees—an *E. coli* genome acquires some sixty new genes, all of them through horizontal gene transfer.[23] And these are only the successful edits—many others have gone out of print and left no descendants.

We already know the DNA sequences of more than a thousand bacterial species, and they testify that *E. coli* is not an exception, but the rule.[24] Most bacterial genomes are just as packed with genes trafficked from other sources, many with an unknown origin, though this is scarcely surprising. Trying to discover the provenance of a particular gene is a bit like trying to trace the literary influences revealed in a single paragraph of a novel by reading a small and random selection of the Library of Congress. A thousand species—or even a hundred thousand—would still be a drop in the ocean of bacterial diversity with countless millions of bacterial species, most of them unknown, all of them potential gene donors.

Not all of this genomic change causes metabolic change, because only a third of a genome is devoted to metabolism—the proteins it encodes also have a lot of other business, helping a cell move around, transporting building materials, and so on.[25] So what if gene transfer mostly shuffles the nonmetabolic parts of the genome? Then evolution's journey through the metabolic library might not take it far, and most metabolisms would therefore be very similar.

Are they? A few years ago, I asked this question of hundreds of bacterial species with known genomic DNA sequences, in a study that relied on decades of research done before my time. This research had discovered thousands of genes that code for particular enzymes, and allowed us to draw maps that connect genes with enzymes, and enzymes with chemical reactions.[26] In other words, we could translate a genome sequence into a metabolic genotype, and compare these genotypes among organisms.[27] And that is what I did.

Figure 5 shows how easy it is to compare two metabolic texts, using the simple example of two short snippets corresponding to ten enzymes in two organisms. Four of the ten enzymes cannot be made by either organism (gray zeroes), six are encoded by the first organism—its genotype string has six ones—and five are encoded by the second organism.

Genotype 1   0110111100
Genotype 2   0100111100

Distance: $D = 1/6$

FIGURE 5. Genotype distance

We take the number of enzymes (six) made by at least one of the two organisms, and the number of enzymes made by only one but not the other organism (one enzyme), and calculate the ratio (1/6) of these numbers. If this ratio were zero, then two organisms would encode exactly the same enzymes. If it were 1/2, then half of the enzymes that one organism can produce would also be produced by the other organism. If the ratio were equal to 1, then the first organism wouldn't be able to produce a single enzyme that the second organism can produce—their metabolisms would be maximally different. Those ratios, ranging from 0 to 1, reflect the difference between the enzymatic portfolios of two organisms, but that's a little unwieldy to write over and over again—better to replace it with the symbol D for *difference* or *distance*. [28]

Unspeakably tedious as comparing genotypes for hundreds of bacteria—each encoding more than a thousand reactions—would be on pencil and paper, my trusted computer can finish it in the blink of an eye. When I asked it to calculate D for hundreds of pairs of bacteria, I was surprised to see—although their highly diverse genomes should have warned me—that even closely related organisms had highly diverse metabolic texts. Thirteen different strains of *E. coli* differed in more than 20 percent of their enzymes.[29] An average pair of microbes differed in more than half of them.[30] I had also suspected that bacteria living in the same environment—the soil, for example, or the ocean—might encounter similar nutrients and thus have similar metabolic texts. Wrong. Their metabolic texts were just as diverse, with a *D* just as different as that from bacteria living in different environments.

This exercise underscores the staggering scale at which nature experiments through gene shuffling. Everywhere on this planet, a relentless shuffling and mixing and recombining of genes takes place. Wherever microbial life occurs, in the depth of the oceans and on arid mountaintops, in scalding hot springs and on frigid glaciers, in fertile soils and desiccated deserts, inside and around our bodies, life is experimenting with every conceivable combination of new genes, rereading, editing, and rejuggling its metabolic texts without pause, yielding an enormous and still growing diversity of metabolisms.

Without readers, a book is a bundle of cellulose sheets with meaningless ink stains. Likewise, a text in the metabolic library needs to be read to reveal its meaning: the metabolic phenotype that determines which fuels an organism can use, and which molecules it can manufacture. We think of a phenotype as something we can see, and many metabolic phenotypes are plain as daylight. They include the melanins that protect our skin against radiation, that camouflage a lion's fur, and that color the ink of an octopus. All of them are molecules synthesized by metabolism. And so are the various pigments that dye tree leaves, lobsters, flowers, and chameleons, whether for defense, courtship, or sometimes for no good reason at all.[31] But metabolic phenotypes do not end at this visible surface. They extend to depths that are hidden from our eyes yet visible to chemical instruments—and to natural selection. Their most important role is to ensure viability itself, which boils down to the ability to synthesize sixty-odd molecules very different from those pretty pigments—they are the essential biomass molecules I mentioned in chapter 2. Viability, viewed as the phenotypic meaning of a genotypic text, is like the simple moral of a complex story, or like a brutally straightforward court judgment: If you can't make all essential biomass molecules, your sentence is death, and it is carried out immediately. Organisms with a mutation that has compromised the ability to synthesize essential molecules don't just fail to live long enough to reproduce. They don't live at all.

To grasp this phenotypic meaning—viability or death—we need to read an organism's metabolic genotype. This is a tall order, not only because the meaning of a text is so much more complex than the text itself—to understand the moral, we have to understand the whole story—but also because our brains are not well practiced in reading chemical language. Fortunately, we can program the artificial intelligence of computers to assist us.

A genotype tells us which reactions a metabolism can catalyze, the molecules these reactions consume, and the molecules they produce. To decipher its meaning, we would first need to know which nutrients are available—without the right ingredients, you cannot bake a cake—and whether the metabolism can use them to build an essential biomass molecule such as tryptophan. This is easiest for the austere minimal environments where survivalists like *E. coli* can thrive, because they contain so few nutrients, sometimes only a single sugar that provides all the carbon and energy the organism needs.

Starting from the available nutrients, we would then write a list of all the molecules the metabolism's reactions *produce* from the available nutrients, find the reactions in the genotype that *consume* these product molecules, and list *their* products, iterating in this way until we find one or more reactions whose products include tryptophan. If no such reaction exists, then the metabolism cannot produce tryptophan.

We would then move on to another biomass molecule, perhaps another amino acid, or one of the four DNA building blocks, repeating the entire procedure for each of the building blocks to find out whether the metabolism can manufacture it. Only when *all* essential biomass molecules can be produced is it viable.[32]

All of this is done on computers, because computation—done right—is faster, cheaper, and can even be better than experimentation. But as the saying goes, the map is not the territory, and we biologists do not fully trust any computation until we can check it. So like a factory that spot-checks its output randomly, we expose organisms with known metabolic genotypes to known chemical environments, and wait, somewhat ghoulishly, for them to grow or die. This has been done, for example, to several hundred mutant *E. coli* strains, each of them engineered to lack one enzyme, and it shows that their computed viability is highly accurate—it is correct for more than 90 percent of strains.[33]

Most biologists who know about this computation think of it as ordinary and do not dwell on how remarkable it is. But more than just remarkable, the capacity to compute viability is profound and revolutionary, a legacy of a hundred years of research in biology and computer science. Darwin and generations of biologists after him could not even dream of it, yet it is crucial to understanding metabolic innovability—nature's ability to create new metabolic phenotypes.

This computation works for any organism whose metabolism we know, and for any known chemical environment, whether Arctic soil, tropical rain forest, oceanic abyss, or a mountain meadow. It also applies to any aspect of a metabolic phenotype—to any molecule a metabolism *could* make. But among all these aspects, viability is the most fundamental, and new methods of making biomass and using chemical fuels are by far the most important innovations. They are also the most far-reaching, opening new territories to life and its metabolic engines.

The reason for the importance of fuel innovations is simple: The world changes all the time, and no matter how successful a metabolism is *today*, it will almost certainly become unsuccessful at some point in the future, like an economy that depends on exhaustible fossil fuels. Chemical environments always change as consumed nutrients ebb and new foods flow. Organisms that depend on a single, specific combination

Fuel Molecules	Phenotype
Glucose	1
Ethanol	0
•	•
•	•
•	•
Sucrose	0
Fructose	1
•	•
•	•
•	•
Citrate	1
Acetate	1

FIGURE 6. Metabolic phenotypes

of nutrients are evolutionary dead ends, and ongoing innovation is needed to survive.[34] Fortunately, many different kinds of molecules *can* provide energy and chemical elements like carbon. Some are as familiar as glucose and sucrose, others less so, like the poison pentachlorophenol.

Even a modest number of potential fuel molecules gives rise to an astounding number of fuel combinations on which a metabolism may or may not be viable—an astounding number of metabolic phenotypes. To see how many, imagine a list like that shown in figure 6, comprising a hundred or so potential fuels. Then compute whether the known metabolism of your favorite animal, plant, or bacterium is viable on a specific fuel molecule, such as glucose. If it can synthesize all biomass molecules from the carbon in glucose, write a "1" next to glucose, otherwise write a "0." Then repeat this computation for the next fuel molecule, the next one after that, and so on, until each fuel has either a "0" or a "1" next to it. Every single "1" in this list means that the metabolism can synthesize the complete suite of biomass molecules from that particular fuel.

The resulting string of a hundred ones and zeroes encapsulates the fuel molecules that a given metabolism can use to sustain life. It is an extremely compact way of summarizing a metabolic phenotype. Metabolic generalists like *E. coli* can survive on dozens of carbon sources, and their phenotype string contains many ones.[35] Metabolic specialists can live on only a few carbon sources, and their phenotypes contain mostly zeroes.

To count how many such phenotypes exist, the different combinations of a hundred-odd fuels on which an organism *could* be viable, we just need to keep in mind that an organism may (1) or may not (0) be able to live on each fuel—these two and no other possibilities exist. To calculate the total number of possible phenotypes, multiply 2 by itself a hundred times, which yields 2^{100}. This number is greater than 10^{30}, or a 1 with 30 zeroes added, not quite as large as the number of possible metabolisms, but still a very large number, much larger than, say, the number of stars in our galaxy—approximately 10^{11}, or 100 billion.

I was not kidding when I told you that phenotypes are more complex than the modern synthesis would have you believe.

This huge number of phenotypes implies an equally huge number of metabolic innovations. Figure 7 shows one example. The figure's left side displays the fuel phenotype of a metabolism that can survive on some carbon sources, but not on ethanol, hence the zero next to ethanol. New genes—acquired through gene transfer or otherwise—can change the genotype that brings forth this phenotype. If this change allows the mutant to metabolize ethanol, we replace the "0" next to ethanol with a "1." Because every conceivable metabolic innovation can be written like this, by replacing a "0" with a "1" in a metabolic phenotype, there are about as many possible metabolic innovations as there are phenotypes.[36]

Designing a space to house the library of all possible metabolisms would be challenging, in part because its volumes exceed the number of

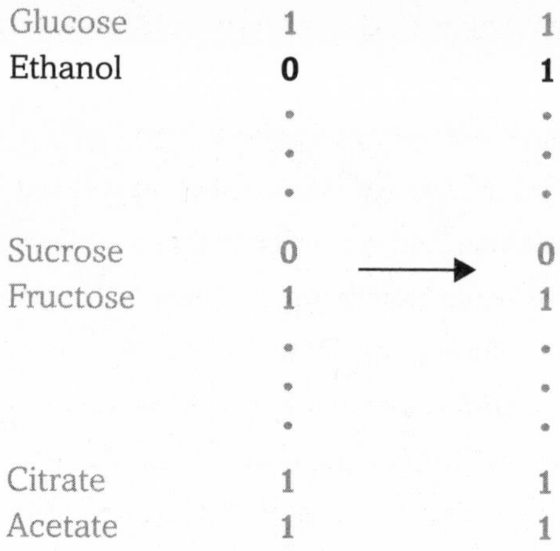

Glucose	1	1
Ethanol	**0**	**1**
	.	.
	.	.
	.	.
Sucrose	0	0
Fructose	1	1
	.	.
	.	.
	.	.
Citrate	1	1
Acetate	1	1

FIGURE 7. A metabolic innovation

hydrogen atoms in the universe. To allow us to find specific volumes fast, the library would also have to be supremely well organized. It would take me only seconds to find my copy of Darwin's *Origin* in the small library of my office, but searching for any one book while grazing through the stacks of an average university's library would be a bad idea. And if somebody had reshelved the *Origin* in the wrong place, it might be lost forever. The problem is much worse in a hyperastronomical library. The universal library might well contain the secret to immortality—or at least the perfect recipe for turkey stuffing—yet the library is so large that we would *never* find it unless we knew where to look.

An especially simple way to organize the library is to place the most similar texts next to each other. Human librarians do exactly that when they shelve different editions of the same book together. If the metabolic library were organized along these lines, the most similar texts would be immediate neighbors. But there is a problem: To buy or build shelving for this library would be a real pain.

In a human library, every book has two immediate neighbors, one to the left and one to the right, or maximally four, if you want to count the volumes on the shelf above and below as well. How many neighbors would any one text in the metabolic library have? Recall that a string of five-thousand-odd ones and zeroes describes a metabolic genotype. Any neighbor would differ in exactly one of these letters, one chemical reaction that may be either present or absent. (It cannot possibly differ in less than that, and if it differed in more, it would no longer be a neighbor.) There is one neighbor that differs in the first letter of this string, another that differs in the second letter, one that differs in the third letter, and so on, until the very last of these letters. In other words, each metabolic text has not two, not four, but thousands of neighbors, as many as there are biochemical reactions, each of these neighbors differing in a single letter and reaction. Shelves that can hold this sort of inventory aren't easy to find.

To see how peculiar they would have to be, imagine a much simpler world than ours, the simplest possible chemical world with only one chemical reaction. In this world the metabolic library has only two texts. One of them consists of the letter 1, containing the one and only reaction, the other of the letter 0—it lacks this reaction. Figure 8a shows these texts as the endpoints of a straight line.

A slightly larger universe with two reactions would be big enough for $2 \times 2 = 4$ possible metabolic texts. One of them has both of these reactions (11), two of them have one reaction but not the other (10, 01), and the fourth metabolism has no reaction (00). Figure 8b shows these metabolisms as the corners of a square.

You may already see where this is going. The next larger reaction universe would have three reactions and $2 \times 2 \times 2 = 8$ possible metabolisms that form the corners of a cube (figure 8c). For a universe with four reactions, we have $2 \times 2 \times 2 \times 2 = 16$ possible metabolisms. But which

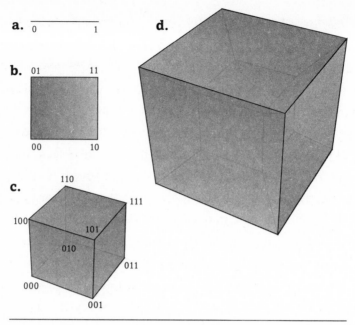

FIGURE 8. Hypercubes

geometric object would correspond to it? As our reaction universe increased from one to two to three reactions, its metabolic texts occupied the endpoints of a line, a square, or a cube, which exist in a one-, two-, and three-dimensional space. Taking it one step further, we need an object in a four-dimensional space. Spaces with four or more dimensions are hard to visualize, but mathematicians routinely work with them, because we can extend our geometrical laws to them.[37] Just as in a square and a cube, the edges of the object we are looking for have to be equally long, and adjacent edges would have right angles to one another. Such an object is a four-dimensional *hypercube*. Figure 8d uses a geometric trick to show this hypercube on paper. It has sixteen corners, each one corresponding to one metabolic text—from 0000 to 1111—that is no longer shown in the figure.

This trick no longer works in five dimensions, much less higher ones. But although it is hopeless to imagine higher-dimensional spaces, they follow the same laws as our three-dimensional space: The edges of a hypercube are equally long, adjacent edges are at right angles to one another, and each corner corresponds to a possible metabolism. And such cubes in high-dimensional space turn out to have curious properties well suited to house the metabolic library.

The number of corners in a square is four, in a cube it doubles to eight, and in a four-dimensional hypercube it doubles again to sixteen. With every added dimension, it doubles, and by the time you have reached 5,000 dimensions, this number has become the hyperastronomical 2^{5000}, the size of the metabolic library. In other words, we can arrange the library's metabolic texts on the corners of a hypercube in a 5,000-dimensional space. This is why off-the-shelf shelving would not work. You cannot cram the metabolic library into three puny dimensions. It needs thousands of dimensions to breathe.

A hypercube is also well suited to accommodate the thousands of neighbors near each of the library's texts. In a simple universe of three reactions, each of the library's texts—the corner of a cube—has three adjacent corners as its neighbors. Take one of these texts, such as the string 100 in figure 8c, and you reach its neighbors via the edges leading from 100 to the adjacent corners. We get to them either by adding the third reaction to 100, which yields 101, or by adding the second reaction (110), or by eliminating the first reaction (000). All three neighbors—101, 110, and 000—differ from 100 in exactly one character. And what holds for one corner of the cube holds for any other corner: It has three neighbors. Likewise, in a 5,000-dimensional cube, each and every metabolism has as many neighbors as there are dimensions, five thousand in all. You can walk from each metabolic text in five thousand different directions, to find one of its five thousand neighbors in a single step. Each of these neighbors differs from the text in exactly one reaction. Either the neighbor has an additional reaction—in this case one entry of the string changes from 0 to 1—or it has one fewer reaction—one entry changes from 1 to 0.

Evolving organisms are like visitors to the metabolic library. Gene deletions and gene transfer allow them to walk through the library, to step from one metabolic text to another, often an immediate neighbor. All of a text's neighbors form a *neighborhood* in this library, and such neighborhoods are as important for evolution as a city neighborhood is for people's lives. City neighborhoods are useful because of proximity—everything is reachable within a few easy steps—and neighborhoods in the metabolic library are important for the same reason. Evolution can reach them in a few small steps, minor edits in a genotype. But residents of a city's neighborhoods can walk in only four cardinal directions—north, south, east, or west—whereas evolution can head in five thousand directions. (Don't even bother trying to visualize that.) And therefore the neighborhood of a metabolic text may be vastly more interesting, surprising, and diverse. This diversity will be crucial to understanding innovability, as we shall see shortly.

Over time, as alterations in an organism's genotypic text accumulate, it walks farther and farther, to more and more distant shelves in the library. To gauge how far, we must be able to measure distance. Without that ability, we would be lost, and the library would become a useless maze of stacks—we could not find our way from one shelf to another.[38] Fortunately, the distance D that I had used to study the diversity of known metabolic texts does the job. It tells us how far apart in the library two metabolic texts are, and it already told us that some viable texts are very distant indeed. The next insight it provides is the real bombshell, though: We can travel enormous distances through the library and encounter very different stories with the same moral, everywhere.

One day we may know millions of metabolic texts, but even that number would be a tiny fraction of the hyperastronomical metabolic library, less than a few specks of dust in the universe, because the library contains many more metabolisms than the number of organisms that have existed on earth since life began. Even after 3.8 billion years of evolution, life has explored only a tiny fraction of the library.

For all of those billions of years, nature did not need to know what was around the next corner of the library for evolution to proceed. But if we humans want to *understand* the library, rather than simply live in it, we need to have some way to grasp where new and meaningful texts are. And we need a catalog that classifies texts, like the Dewey Decimal System, or the Library of Congress Classification, grouping books according to subject categories—Art History, Economics, Linguistics—with smaller subcategories such as Romance, Germanic, Slavic languages nested within them. Metabolic phenotypes, the possible meanings of a metabolic text, are the natural subject categories of this library. Their number is larger than those in a library of books, but that's simply because the library itself is so vast.

A catalog is like a map for this library—it is a genotype-phenotype map that tells us where to find the genotypes with any one phenotype. Without this map, we do not know whether texts with the same subject are scattered or grouped—as they would be in a human library—whether the same shelf houses texts on different subjects, and so on. And because no librarian is in sight, we need to create this map ourselves, roam the library and explore it, like the ancient voyagers who mapped the earth and its continents on their journeys. The library's huge size will prevent us from mapping every single text, but we can draw the contours of the continents, mountain ranges, rivers, lakes, and deserts, and hope that we can grasp the shape of the whole from their hazy outlines.

But where to start, and how to travel?

Here is a puzzle that will point the way. Take a metabolism with any one phenotype, such as viability on glucose, and ask, What if only one text in our library of more than 10^{1500} metabolisms expressed its meaning? As many as five nonillion (5×10^{30}) bacteria exist on earth today.

This number is vast, a 1 with more than 30 zeroes. But even if each of these bacteria had tried a new enzyme combination every second since life began almost four billion years ago, they would have tried only about 10^{48} such combinations.[39] Their chances of having found the one and only working combination would be vanishingly small, smaller than one in 10^{1450}. This number—so small as to be effectively meaningless—means that it would be utterly impossible to find this text through a blind search.

On the one hand, the odds against finding just *one* useful metabolism are vast. On the other hand, life's diversity shows that evolution had no problem finding it. This means that our premise must be wrong: There has to be more than one metabolism—perhaps even many—that solves the problem of surviving on glucose.

To find them, let's do what evolution does: journey through the library and edit genomes—through a series of gene transfer or deletion events that add or eliminate at least one gene, enzyme, or reaction. The starting point for such a journey isn't terribly important. It could be any text in the library, any text that encodes a metabolism viable on glucose or on any other fuel.

So let's start with a metabolism viable on glucose, and either delete a randomly chosen reaction or add a randomly chosen reaction from the known reaction universe. Nature would make a simple and brutal evaluation of the new text: life or death. But we scientist travelers are privileged, because we can retrace our steps. We compute the meaning of the altered text, and, if it turns out not to be viable on glucose, return to the starting text, and add or delete another random reaction—remember, there are five thousand ways of doing that. But if the neighbor is viable on glucose, the journey continues. We add or delete a second reaction, compute the phenotype, and repeat, more or less ad infinitum.

In other words, step from a starting text to its neighbor, to the neighbor's neighbor, to the neighbor's neighbor's neighbor, and see how far you could walk without ever changing its chemical meaning, viability on glucose. Because each step alters a text at random, this walk is a *random walk* through the metabolic library, similar to how a drunkard might stagger home from a night out at the bar, with one difference: Each step in our random walk must encounter a text with the same meaning, the same phenotype.

If there were only one metabolism viable on glucose, this random walk would lead literally nowhere, because the starting text would have no viable neighbors. We would be rooted to the spot. The same would be true if there were a few such texts scattered widely through the library—we could not reach them without destroying viability on the way. And even if they were close together the random walk might not lead far. A few neighbors of the starting text might be viable, but *their* neighbors might not be.

Only if many such texts existed could we roam the library. But in that case we would face a different problem altogether: computing power. To compute one text's meaning is a breeze, but what if this random walk had thousands of steps, and each could lead in thousands of different directions. This is the sort of problem that could take an off-the-shelf desktop computer years or decades to solve. An entire network of computers—a computing cluster—is required to speed up that computation. And that costs money.

While I was slowly advancing from a Ph.D. student to a postdoctoral researcher, and eventually to a tenured professor at a U.S. research university, funding for the kind of basic research that addresses the problem of evolutionary innovation began drying up. This drought combined with the ailing health of my European family, so when a job offer arrived from Switzerland, I was ready to take a leap across the Atlantic, back to my European roots.

I knew that Switzerland was a world leader in science, enormously productive, and technologically sophisticated.[40] Its world-class system of public education, generous support for academic research, and attractive living conditions are behind this success. I would be sad to leave many dear academic colleagues behind, but the opportunity to join the Swiss scientific community was a privilege both humbling and enticing. Most important, the offer was good enough to finance not only a computing cluster but also a state-of-the-art experimental laboratory. Even better, it would allow me to recruit multiple like-minded researchers from all over the world. It was an offer I could not refuse.

On a crisp fall day in 2006 at the University of Zürich, I was sitting in my newly furnished office, inside an austerely elegant building whose simple geometric contours are drawn in a gleaming blend of glass and metal, when a young Portuguese man walked in. Handsome, soft-spoken, with curious deep brown eyes and a quick smile, he introduced himself as João Rodrigues.

João had studied physics, but he had heard that there were many exciting problems waiting to be solved in biology. He was looking for a new challenge, a difficult problem to crack that would get him a Ph.D. He did not know much biology at the time, but he had assets that many biologists lacked: He was good at mathematics, knew how to program computers, and had already performed large and complicated computations. When I first saw his résumé, I could hardly contain my excitement. João had exactly the skills needed to navigate the vast metabolic library. During his job interview, I shared my passion about learning how nature creates. Fortunately for me, we connected. His eyes lit up. He signed on.

João's background is typical for researchers in my lab. They hail from a dozen different countries in the Americas, Europe, Asia, and Australia, and from many disciplines, including biology, chemistry, physics, and mathematics. This is not a coincidence, because the problems we tackle require new skill combinations, so much so that I like to compare our work to that of evolution: Studying innovations, like creating them, benefits enormously from novel combinations—not of enzymatic but of intellectual skills.

I soon became impressed with João's computing wizardry, even though I remained worried that the cluster of more than one hundred computers we had built would still be too slow, that we would never leave the first shelf of the library. But João tricked the machines into working faster, accelerated their computations many times, and eventually launched us far into the library's vast stacks.

João's exploration started with a single well-studied metabolism, that of the bacterium *E. coli* and its viability on glucose—the ability to synthesize all its sixty-odd essential biomass molecules from this single sugar.[41] To find out whether only one metabolism with this ability existed, João first created more than a thousand of *E. coli*'s neighbors, each of them a metabolism that differs in a single chemical reaction from *E. coli*. If *E. coli*'s metabolism is an instruction manual to make all essential biomass molecules, then these neighbors are minor variations on the manual. The first question: Do any of them contain sufficient information to produce all sixty biomass building blocks from glucose?

João computed the answer and quickly found that not one, not two, not three, but hundreds of *E. coli*'s neighbors are viable on glucose. This discovery contained a simple but vital lesson: The uniqueness of this phenotype is but a deeply flawed prejudice.[42] The neighborhood of any one text contains many other viable texts like it. But nothing had prepared us for what came next, when we began to venture further.

João used *E. coli* as a starting point for deep probes of the metabolic library that led further and further away from the starting text. The objective was to learn how far we could travel—hopping from one viable text to a viable neighbor, to the neighbor's neighbor, and so on—without losing viability on glucose. How radically could a metabolic text be edited without losing its meaning? When João showed me the answer, my first reaction was disbelief. The furthest viable metabolism he found—the one with the highest D—shared only 20 percent of its reactions with *E. coli*. We had walked, computationally speaking, almost all the way through the library—80 percent of the distance that separates the furthest volumes—before we were finally unable to find a glucose-viable text by taking a single step.

Worried that this might be a fluke, I asked João for many more random walks, a thousand more, each preserving metabolic meaning, each leading as far as possible, each leaving in a different direction—possible only because the library has so many dimensions. When the answer came back, I was stunned once again. These random walks had led just as far away as the first one. Each of them led to a metabolism that differed in almost 80 percent of its reactions from *E. coli*. They had found a thousand metabolic texts that shared very little with *E. coli*, except that all of them could produce everything a cell needs from the carbon and energy stored in glucose. If we had kept on walking, we would have found even more texts, too many even to count, although later we were able to estimate their number in parts of the library.[43] For example, the number of metabolisms with two thousand reactions that are viable on glucose exceeds 10^{750}.

The number of texts with the same meaning is itself hyperastronomical. The metabolic library is packed to its rafters with books that tell the same story in different ways.

While we surely had not expected this, our explorations had revealed an even more bizarre feature of the library. The thousand random walks did not end in a few stacks of the library, where texts with similar meanings might huddle in small groups—groups of metabolisms with similar sets of reactions. These texts were just as different from each other as they were from that of *E. coli*—they encoded metabolisms with very different sets of chemical reactions. The library does not have clearly distinct

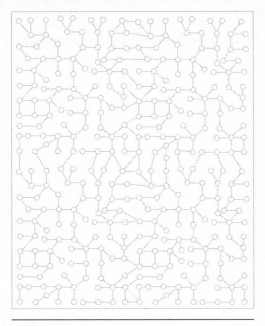

FIGURE 9. A genotype network

sections, like rooms that separate all texts on history from those on science.[44]

And even more surprising was what we found when we started new random walks from these texts—as we had done from *E. coli*—and walked toward other texts without ever changing viability. We always succeeded in reaching them, no matter how far away they were from our starting point. Every single time. This taught us that a connected network of paths linking texts with the same meaning extends throughout the library. I call this network a *genotype network*. It might look a bit like the network of straight lines in figure 9, where the large rectangle stands for the metabolic library, and the lines connect neighboring texts (circles) with the same meaning. Pictures like this are wobbly visual crutches—two dimensions instead of five thousand, a handful of texts instead of unimaginably many—but they are all we have to visualize places as strange as this.

In an ordinary public library, you might find biographical

information about Charles Darwin in one text on a shelf in the history section, and another in biography. In a large research library using the Library of Congress's classification system, you might find some such texts in section QH (for "Science: Natural History, Biology"), but others in sections DA ("World History, Great Britain"), GN ("Anthropology"), PR ("English Literature"), and even BL ("Religion, Mythology, Rationalism"). But you would find nothing that resembles the organizing principles of the metabolic library. You would not find a network of meaning-preserving paths connecting the Darwin biography in HM ("Sociology, General") with another in BT ("Doctrinal Theory"). You would not be able to walk from one book to its neighbor, to the neighbor's neighbor, and so on, almost all the way through the library, without ever being farther than one book away from another that told Darwin's life story in different words.

In the metabolic library, though, that's exactly what a browser can do. Its myriad texts with the same meaning could be like stars in our universe, islands separated by vast expanses of dark empty space. But they are not. You can travel between them on a network of well-lit paths.

Thus far, we had cataloged only the volumes of one subject area— viability on glucose—but many other subject areas exist. There are metabolisms that are viable on ethanol, acetate, and dozens of other fuels. And we mapped them, using the same random browsing strategy: different random walks whose editing steps preserved the phenotype— viability on ethanol, for example—and stopped only when we could walk no further. We did that for eighty different fuels, and each time we saw the same pattern. Viable metabolisms can have very different texts—they share as little as 20 percent of their reactions—and form a vast connected genotype network in the metabolic library.

Emboldened by this general pattern, we began to map metabolisms viable on multiple different fuels like ethanol, glucose, and acetate, able to synthesize all biomass molecules from each of them. (The advantage of this ability is obvious: It permits survival when the supply of any one fuel runs out.) Because that metabolic skill would be more difficult to achieve, perhaps only a few metabolisms might have it, all of them shelved in one corner of the library? We were again proven wrong. We studied metabolisms viable on five, ten, twenty, and up to sixty different fuel molecules. Each meaning-preserving random walk starting from one of them led far away. Even some metabolisms viable on sixty different fuels shared fewer than 30 percent of reactions. And the metabolisms with the same phenotype—countless trillions in each subject category—again formed a connected genotype network.[45]

At this point, I was close to ecstatic. We had stumbled upon fundamental principles that govern the metabolic library's organization. First, many metabolisms are viable on the same fuel molecules—it matters little which fuels you choose. Organisms can assemble biomass building blocks in many ways, through many different sequences of reactions. Second, many of these metabolisms are very different from each other, sharing only a minority of reactions. Third, the viable metabolisms we found were connected in a gigantic network—a genotype network. This genotype network reached far through the space of metabolism.[46] Each subject area has one such genotype network, and these networks form a densely woven fabric in the metabolic library.

We had accomplished all this with modest means, because our computing power was puny when compared to the number of texts in the metabolic library. We had crudely mapped a world vast beyond imagination. We had crossed an ocean in a bathtub.

Myriad metabolic texts with the same meaning raise the odds of finding any one of them—myriad-fold. Even better, evolution does not just explore the metabolic library like a single casual browser. It crowdsources, employing huge populations of organisms that scour the library for new texts.

Every time gene transfer alters the metabolic genotype of an organism, it takes a step through the library. Different readers—billions of them—walk off in different directions to explore the library.

Evolution's library exploration also differs in another way from how we humans would browse a library. Imagine a hapless organism that steps off the path connecting viable metabolisms by encountering a change—perhaps a gene deletion—that disrupts the metabolic instructions for manufacturing a key molecule. That organism would be dead, courtesy of natural selection. In the metabolic library readers die (and others get born) in an exploration that unfolds over generations.

Viewed from afar, the library's explorers, from bacteria to blue whales, might appear like giant clouds of dust grains—dwarfed by the library itself—drifting this way and that, from one stack to the next, endlessly meandering swirls of living things that try new combinations of chemical reactions over and over and over again. Some die. Others survive, and pass innovative combinations on to subsequent generations. This churning mass of life is evolution in action.

That action would vanish if genotype networks did not exist. If only one text could confer viability on any one fuel, then all members of a population would have to share that text, crowding around it in the library. Whenever a member stepped aside to sneak a peak at another volume, it would die. If only a few similar texts were viable, the population could only explore a tiny section of the library. But because of genotype networks, evolving populations can explore the library far and wide.

Genotype networks are the first of two keys to innovability. And now for the second: the immense diversity of the neighborhoods where these explorations begin.

Imagine a patch of soil where billions of bacteria can thrive as long as new food arrives occasionally—a leaf blown in from afar, a rotting carcass, or perhaps a ripe apple dropping from a tree. Many molecules in these foods are nutritious, but they would be useless unless one of the microbes had acquired the right combination of enzymes, the right metabolic text for transforming the food into biomass. That very text could be lifesaving once the billions of its soil-mates, all of them hungry, have consumed and exhausted other fuel molecules. This text, a metabolic innovation, could give one microbe a new lease on life.

Even if there were only a hundred fuel molecules, some 10^{30} metabolic phenotypes would exist, and finding this text is to find a specific one of them. There is simply no way you could pack 10^{30} texts into a small neighborhood of the library. Each neighborhood only has enough space for a few thousand texts, whose meanings can encompass a tiny fraction—one in 10^{26}—of all possible phenotypes. It's as if you borrowed a few volumes at random from the New York Public Library to fill your nightstand and hoped to find *The Origin of Species* among them—almost impossible. But these odds change if a crowd of readers can browse the library along a genotype network that extends far through the library. Because genotype networks are so large, the population could explore thousands of neighborhoods and increase the odds of finding the new lifesaving phenotype.

You may have noticed a hidden premise here: Different neighborhoods must contain different novel phenotypes. Near a volume on photovoltaics, you would find others on medieval French literature, twentieth-century architecture, and Italian cooking, whereas near *another* volume on photovoltaics—in a different part of the library— books on toy trains, World War II, and astrophysics would be shelved. In metabolic terms, you might find metabolisms viable on acetate and ethanol and citrate in one neighborhood, and metabolisms viable on sucrose and fructose in another.

To find out whether this bizarre library organization really exists, we chose pairs of metabolic texts that had the same phenotype (viability on glucose) but that were otherwise very different. The two metabolisms, A and B, were located in different parts of the library—they did not share many reactions—yet both were part of the same genotype network. We then examined the phenotypes of all their five-thousand-odd neighbors, and found that some of them were likewise viable on glucose—they belonged to the same genotype network—while others had lost a critical chemical reaction, which spells death. Yet other neighbors—those we were really interested in—could live on a new combination of fuels, such as ethanol or fructose. For these networks we asked: Do the neighbors of metabolic genotype A—those texts that differ from A in only a single reaction—contain metabolic innovations different from those of the neighbors of metabolic genotype B? If the neighborhood of A contained metabolisms viable on the new fuels ethanol and fructose, would the neighborhood of B contain metabolisms viable on, say, acetate and sucrose?

After analyzing thousands of network pairs, and after studying phenotypes involving eighty different fuel molecules, we had found that the premise was correct. Different neighborhoods contain texts with new meanings, but these meanings differ between neighborhoods. Most metabolic innovations are unique to one neighborhood and do not occur in the other. (Because each new phenotype has its own genotype network, this also means that different genotype networks in the library are interwoven in an unfathomably complex way.)

We then went one step further. With our computers' help, we wandered once again through a genotype network in the metabolic library, except that now we behaved like inventory clerks with their notepads, listing all the innovations in the immediate neighborhood of our path—all innovations that were within easy reach. We listed all the different new phenotypes in the walker's neighborhood before the start of the walk, and examined the neighborhood again after the first step. If it contained a new phenotype that was not already on the list, we added it to the list, took one further step, examined the new neighborhood, added any new phenotypes, and so on, for thousands of steps. Because we knew that different neighborhoods contain different innovations, we expected the list to grow over time, as new phenotypes became accessible. But we expected that we would run out of new phenotypes eventually.

Wrong. Long after our notepads were full, we were still encountering innovations.

Worried that this trip had yielded an unusually rich bounty, we went on many more trips, from different starting points in the library, metabolisms viable on different fuel molecules. And we also crowdsourced our shopping, exploring the library not with a single metabolism but with entire populations of evolving metabolisms to tally how many different new phenotypes they found. In every instance, innovations continually piled up, with no sign of slowing down, at a steady clip, unceasingly, no matter how long the exploration continued, a hundred, a thousand, or ten thousand steps, hours, days, weeks, until we ran out of time and needed to do other work. We realized that the innovability of an evolving metabolism would not exhaust itself in our lifetime.[47]

Innovability in the metabolic library is near limitless, and for that both genotype networks *and* diverse neighborhoods are required. They are the two keys to innovability. Genotype networks guarantee that evolving populations can explore the library. Without them the lethal punishment of losing viability would be inevitable. But without diverse neighborhoods in this library, exploring a genotype network would be pointless: The exploration would not turn up many texts with new meanings.

Any librarian who wanted to organize a human library in this way would be locked away. Even if a thousand books told the same story in different ways, no sane librarian would create sections that placed books with all manner of different meanings next to one another. And he would certainly not pack *different* neighborhoods around synonymous texts with books in *different* subject categories.

But a closer look reveals that the metabolic library's catalog is far from a madman's febrile fantasy. Human libraries are useful only because we *have* librarians who make catalogs suitable for us, where books on photovoltaics stand on one shelf, those on French literature on another, and so on. For a library whose readers have no catalog and can only take random steps, and where missteps are punishable by death, it would be disastrous, because they would be stuck on whatever shelf they started. They would be idiots savants, world experts in one area but completely ignorant in all others, and could never learn anything new—not a smart strategy for surviving in an ever-changing world. For such readers, the metabolic library is perfect, uncannily well set up for innovation. It guarantees eternal learning and innovability.

Even more uncanny: Life's other libraries are organized the same way.

CHAPTER FOUR

Shapely Beauties

The Arctic cod is a slender, brownish fish, with a silver belly and black fins, between eighteen and thirty centimeters long, a perfectly unremarkable occupant of the world's oceans. Except for one thing: The Arctic cod—*Boreogadus saida*—lives and thrives within six degrees of the North Pole, nine hundred meters beneath the surface, in waters that regularly chill below zero degrees Celsius.

At that temperature, the internal fluids of most organisms turn into ice crystals with edges as beautiful as well-forged swords, and just as deadly, for they carve up living tissue like butter. Warm-blooded animals have a built-in thermostat that allows them to survive in subfreezing weather. Fish don't. And yet, there's the Arctic cod.

B. saida survives by producing antifreeze proteins that lower the freezing temperature of its body fluids, much like the antifreeze in a car's engine coolant. These proteins are prototypical examples of nature's innovative powers. Change the amino acid sequence needed to produce a particular protein and, presto, huge areas of the earth's oceans become livable.[1]

Antifreeze proteins are among thousands of innovative wonders that populate the cells of fish and of every other living being. If you could shrink yourself and travel through a cell, you would first be astonished by how many *different kinds* of molecules there are, millions of them. Tiny molecules like water, larger molecules like sugars or amino acids, and even larger macromolecules like proteins all push and jostle and shove past each other like subway commuters during rush hour.

Proteins, the giant hulking monsters of a cell's molecular population, are life's workhorses. We have met the metabolic enzymes that synthesize everything a cell needs—including their own amino acids—by linking smaller molecules, cleaving them like molecular scissors, or simply rearranging their atoms.[2] But not all proteins are enzymes. Some are molecular motors, like the proteins that help your muscles contract, or like the *kinesins* that "walk" along stiff molecular cables that crisscross the cell, carrying tiny membrane vesicles that shelter various molecular cargoes. Mayhem ensues when these truckers of the cell no longer do their job. One kinesin, for example, transports building materials needed to wire the cells in our nervous system, and mutations in its gene can cause an incurable disease called Type 2A Charcot-Marie-Tooth disease, which hampers movement and sensation in feet and hands.[3]

Yet other proteins attach to DNA and switch genes on or off. These *regulatory* proteins allow the information encoded in a gene to become transformed into an amino acid string. Hundreds of such regulators work simultaneously, each of them flipping the switch on some genes but not on others. (They are the source of yet another kind of innovation we will explore in chapter 5.)

And there's more: rigid protein rods that form a cell's molecular skeleton, proteins that import nutrients, proteins that dump waste outside the cell, proteins that relay molecular messages between cells, and on and on.

Each of these proteins has its own special talent, expressed in its phenotype, whose most important aspect is shape.[4] I do not just mean the molecular shape of the twenty kinds of amino acids in proteins, and the order in which they are strung together—the *primary structure* of a protein.[5] I mean the shape this string forms in space through the protein folding process that I first mentioned in chapter 1.

Hydrophilic amino acids love to be near the water that surrounds them, whereas *hydrophobic* amino acids avoid water—like the oily parts of membrane molecules—and these molecular sympathies help an amino acid string fold in a stereotypical manner. Driven by heat's vibrations, a folding protein explores many shapes of its amino acid chain until it finds one where most water-avoiding amino acids cluster together and form a densely packed core that is surrounded by the water-loving molecules on the protein surface.[6] What's more, some amino acids attract and others repel each other, and these chemical sympathies also influence a protein's fold. The protein folding process—driven by nothing but erratically bouncing molecules—is yet another reminder of the power of self-organization. It occurs millions of times a day in each of our trillion cells, whenever they manufacture a new protein string.

Viewed atom by atom, the three-dimensional *fold* of a protein like the sugar-splitting sucrase from chapter 2 would appear like a shapeless blob. But by stepping back and focusing on the string that holds the amino acid beads together (figure 10), one can discern regular and repeated patterns of amino acid arrangements in space that occur in many proteins. They include helixlike corkscrews—one is labeled in the figure—or several parallel strands, an arrangement also called a *sheet*.[7] Helices and sheets are major elements of protein folds and form the *secondary structure* of a protein. And these helices and sheets, together with the strands that connect them, form the labyrinthine three-dimensional or *tertiary structure* of a protein's fold shown in figure 10.

Even though it may look like a tangled pile of spaghetti, the fold in figure 10 is actually highly organized: Any two sucrase amino acid strings spontaneously fold into the exact same shape.[8] This shape is critical

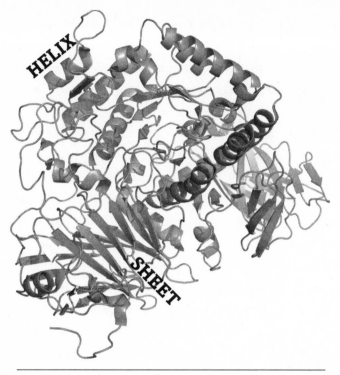

FIGURE 10. Sucrase folded in space

to a protein's function, because its helices and sheets guide and constrain the endless heat-induced vibrations and oscillations of the folded protein. Such constrained movement allows enzymes like sucrase to cleave sucrose—a bit like the blades of scissors whose movement is constrained by the pivot that connects them and enables them to slice paper.[9] Because heat-caused vibrations are so important to enzymes, these molecules also have an optimal temperature: Too little heat and their vibrations are not powerful enough to reorganize molecules. Too much heat and the vibrations shake the fold apart—into the linear string of amino acids. Worse than that, unfolded proteins often aggregate into large inert clumps like the proteins in boiled eggs. Such clumps of unfolded proteins are more than useless: When too many of them accumulate, for example in your brain, they bring forth horrible diseases like Alzheimer's.[10]

In the bewildering realm of oscillating shapes that sucrase and other proteins inhabit, each shape has a specific job. Each is well suited for what it does and highly complex. In the words that Darwin used to describe the living world, it is a world of "endless forms most beautiful,"[11] but *these* forms—unknown to Darwin—keep the living world alive.

Proteins do not just perform existing jobs. The economy of living organisms, like that of the human world, is constantly changing, and in response, evolution brings forth new protein shapes, innovations that take on new jobs. These jobs open whenever life needs to solve a new problem, like that of surviving the menacing knives of growing ice crystals.

And just as in the human economy, where inventions from blast furnaces to smartphones are often made several times independently, the innovations that fill these jobs are often discovered more than once. Antifreeze proteins are a case in point: They originated not just in the Arctic cod but also in Antarctic fish, and from different proteins in their ancestors.[12] They even originated more than once in the Arctic.[13] What's more, some fish evolved more than one kind of antifreeze protein. The winter flounder, a flatfish from the North Atlantic, manufactures one antifreeze protein that prevents its bloodstream from freezing, and another that protects its skin.[14] And some of these proteins arose very quickly in evolutionary terms—in less than three million years.

Dozens of amino acids had to change from proteins in some frost-sensitive ancestor to create antifreeze proteins, but protein innovations often require much less change.[15] Alter as little as one amino acid in the enzyme needed to manufacture the amino acid histidine, and the result is a new enzyme that helps manufacture the amino acid tryptophan.[16] Mutate a specific amino acid in an *E. coli* enzyme that helps extract energy from the sugar arabinose—its name comes from gum arabic, a natural gum from acacia trees—and this enzyme transmogrifies from a rearranger of atoms to a cleaver of molecules.[17]

Such minimal changes can have dramatic consequences for life, as the bar-headed goose from Central Asia could tell us. It is one of the world's highest-flying birds. It has to be, since its migratory route takes it across the Himalayas at altitudes that exceed five miles, where the air is not only thinner—requiring birds to flap harder—but contains only a third as much oxygen by volume as sea-level air. At that altitude, the mountaineers who struggle up Mount Everest use oxygen tanks, and the passengers on jet airliners require pressurized cabins. The goose can't benefit from either technology, but no problem, it has an even better trick. Its hemoglobin—the protein that shuttles oxygen from lungs to muscles—harbors an amino acid change that helps it bind oxygen much more tightly than our hemoglobin. It allows the goose to scavenge oxygen molecules from thin air, and keeps this bird flying where others are grounded.[18]

Molecular innovations like the Arctic cod's antifreeze or the bar-headed goose's oxygen-binding hemoglobin are valuable because they expand an organism's habitat, which means more food, better survival, and more offspring. Other innovations confer a different kind of advantage, such as the ability to discriminate between one kind of food and another, to choose a nutritious rather than a poisonous plant for dinner. They depend on improving perception, rather than mobility, and they are why the retina in the back of our eye contains three kinds of opsins. These are highly specialized proteins that detect light and are tuned to the different wavelengths of blue, red, and green light. Thanks to them, we see the world in color. This was not always the case. Our most distant ancestor among the vertebrates probably had only one opsin. Theirs was a black-and-white world. Most mammals have two different opsins, those for red and blue. They can see in two colors. But we and some of our relatives like chimpanzees can see in three colors, perhaps because color vision helped our distant ancestors forage: It lets fruits stand out from the background of green foliage. Whatever the reason, the innovation of color vision takes very little change, as little as three altered amino acids that retune an opsin from red to green.[19]

Innovations like color vision benefit us, but others harm us—those of deadly bacteria that resist the antibiotics your doctor prescribes. They are the unfortunate side effect of our continual improvement of antibiotics, the result of a biological arms race of bacteria against biotechnologists. This race evokes the Red Queen of Lewis Carroll's *Through the Looking Glass,* who famously told Alice, "Now, *here,* you see, it takes all the running *you* can do, to keep in the same place."[20] Through it, bacteria have discovered various protein innovations, some of which destroy antibiotic molecules, whereas others, known as *efflux* pumps, force antibiotics out of the cell like some bacterial rescue squad pumping toxic gas out of a contaminated house. (Horizontal gene transfer, combined with human travel, can spread such innovations throughout the world within months.) Especially sinister are proteins that pump not just one but many kinds of antibiotics, and thus render bacteria resistant against multiple antibiotics. Curiously, when our own body cells go rogue, proliferate wildly, and evolve rapidly—in cancer—they often use similar efflux pumps to rid themselves of unwanted cancer drugs. These are not only independent solutions to a similar problem but also one of many reasons why the war on cancer is hard to win.[21]

The proteins behind these innovations were not created from scratch. They are modified transporters, proteins that are essential in a cell's daily life, because they ship thousands of molecules—nutrients, waste, building materials—to various destinations within the cell. So should we really call them innovations? The same question arises for the goose's improved hemoglobin and primates' color vision. Nature just fiddled with hemoglobin to tighten its binding to oxygen, and it tinkered with opsins to tune their color sensitivity. Neither was a qualitatively new protein. But consider the impact of these changes. Consider the millions of square miles of new habitat opening up to a bird that can traverse *any* mountain range. Consider how much duller our world would be in black and white. And consider the life-and-death change that drug resistance can make to a bacterium. For their dramatic consequences alone, these small changes deserve to be called innovations. They show how minute alterations of no more than a few atoms can have effects that percolate through an organism that is a million times as large and alter the life of its descendants forever.[22]

In chapter 3, we saw how nature continues to create ever-novel sequences of chemical reactions, by combining and recombining metabolic enzymes through horizontal gene transfer. But that is not how metabolic enzymes themselves first appeared. As the last few examples showed, nature creates new proteins, including every one of the known five-thousand-plus enzymes, by altering the amino acid sequence of their protein ancestors. That's also how it created the countless proteins that regulate genes, ferry materials, contract muscles, transport oxygen, import nutrients, export waste, communicate between cells, and perform a thousand other tasks. Entire books could be written—have been written—that describe a few such innovations in great detail.

This book is not among them.

You cannot understand what made all these innovations possible through anecdotes—an antifreeze protein here, an opsin there—any more than you can draw a map of the Unites States with satellite images of a few counties. The task requires comparing many old proteins and the new ones they brought forth. Thousands of them.

This task is made easier if one can read the DNA of genes or the amino acid strings they encode—the genotypes of proteins.[23] Among the first learning to read both was the British biochemist Frederick Sanger, one of few scientists to win two Nobel Prizes—the first for deciphering the amino acid sequence of insulin, the second for techniques to read the letter sequence of DNA. His discoveries came decades earlier than our ability to read the genotypes of metabolisms, and we therefore know many more protein genotypes and phenotypes.[24] They hail from organisms that live in Arctic wastelands and tropical jungles, on mountaintops and in ocean depths, in our gut and in boiling hot springs, in barren deserts and in fertile soil, in filthy sewers and in pristine rivers.

Without organization, this giant heap of protein facts would be like a million shuffled words in a madman's dictionary, but once organized, it becomes part of a library just like the gigantic metabolic library from chapter 3. The volumes in *this* universal library are protein genotypes, texts written in a twenty-letter alphabet, where each letter corresponds to one amino acid. The universal protein library is the collection of all proteins that life has created, and all proteins that it *could* create. It is sometimes also called a *protein space* or a *sequence space*—because each text corresponds to a single sequence of amino acids.[25]

The size of this library is no less staggering than that of the metabolic library, as an already familiar calculation helps us see. Recall that there are 20 × 20, or 400, possible two-letter texts using one of twenty possible amino acids. Similarly, there are 20 × 20 × 20, or 8,000, texts of three amino acids, 160,000 texts of four amino acids, and so on. Short texts like this are called peptides, but most proteins comprise much longer texts—polypeptides—and the number of possible amino acid texts explodes with their length, such that the number of proteins with merely a hundred amino acids is already greater than a 1 with 130 trailing zeroes. But the library is larger than even this unimaginably large number, because proteins like sucrase have more than a thousand amino acids, and some human proteins are many times longer. (Among them is a behemoth called *titin,* a 30,000-amino-acid-long protein spring in our muscles.)[26] The universal library of proteins is another library of hyperastronomical size.

The similarity to metabolism does not end with the size of this library. Like the metabolic library, the protein library is a high-dimensional cube, with similar texts near one another. Each protein text perches on one vertex of this hypercube, and just like in the metabolic library, each protein has many immediate neighbors, proteins that differ from it in exactly one letter and that occupy adjacent corners of the hypercube.[27] If you wanted to change the first of the amino acids in a protein comprising merely a hundred amino acids, you would have nineteen other amino acids to choose from, yielding nineteen neighbors that differ from the protein in the first amino acid. By the same process, the protein has nineteen neighbors that differ from it in the second amino acid, nineteen neighbors that differ from it in the third, the fourth, the fifth, and all the way through the hundredth amino acid. So all in all, our protein has 100 × 19 or 1,900 immediate neighbors. A neighborhood like this is already large, and it would be even larger if you changed not one but two or more amino acids. Clearly, this can't be bad for innovation: With one or a few amino acid changes, evolution can explore many proteins.

In another parallel to the metabolic library, you would get lost wandering through *this* library's maze unless you had an unrolling skein of wool to gauge how far you traveled. Once again, a notion of *distance* serves this purpose. It is the number of amino acids by which two proteins differ. It tells you how far you need to walk—how many amino acids you need to change—to travel from one protein text to any other.[28]

The texts in this library are important, but even more important is the meaning each one carries. Our eyes cannot read this meaning, the words, sentences, and paragraphs of a protein's chemical language, but life is fluent in this language. And it can tell whether a protein is meaningful or embodies jumbled chemical ramblings.

Cells take a hard-nosed view on which proteins are meaningful: those that help them live. A protein is meaningful only if it is useful, and defective mutant proteins that do not fold properly have lost their meaning. If a protein's "meaning" feels too anthropocentric a word, it is worth reminding ourselves how "meaning" is defined by semiotics, an offshoot of linguistics that explores the meaning of meaning: whatever a sign—which could be anything from a road sign to a book's text—points to. If that sign is a protein's amino acid text, then the meaning it encodes is the protein's phenotype and the function it serves inside a cell.[29]

We still do not know how many meaningful books a universal library of books would contain, but decades of research allow us to estimate this number for proteins, because most useful proteins fold into a specific shape. If you blindly took a random protein from a random shelf in the library, the odds that it folds are at least one in ten thousand. That may not seem much, but keep in mind how vast the library is, containing more than 10^{130} proteins of a hundred amino acids. Even if only one in ten thousand of them folds, you are still left with 10^{126} proteins, a 1 trailed by 126 zeroes, much greater than the number of hydrogen atoms in the universe. The number of meaningful proteins is itself large beyond imagination.[30]

Evolution explores the protein library through huge populations of organisms. Their proteins change, one amino acid at a time, with the occasional copying errors that alter a DNA string's letters—A to C, T to G, or in any other way—as this string replicates generation after generation. To understand how such change creates texts with new and useful meaning, we need to map the protein library like we mapped the metabolic library. This is less difficult than it seems: Thanks to decades of work by armies of protein scientists, we know the folds and functions of tens of thousands of proteins and their place in the library. What's more, the technologies of twentieth-century molecular biology allow us to take any volume off its shelf—to manufacture any protein—and study its fold and function in the laboratory.

The simplest question about innovability in proteins is one we encountered before. How hard is it to find a protein with any one meaning, one whose function helps an organism to survive? If there is only one of it in the library, even the eons elapsed since the Big Bang would not suffice to find it. Since meaningful proteins exist in huge numbers, just about every problem that life solved with a protein innovation must have more than one solution. But how many?

In 2001, Anthony Keefe and Jack Szostak from Harvard University set out to answer this question for a family of proteins whose invention was as crucial as any in life's history: the proteins that can bind the ATP that we encountered as life's battery in chapter 2. Proteins that carry out work—they transport materials, contract muscles, build new molecules—cleave ATP, and in doing so, harness its energy for this work.[31]

To use ATP's energy, a protein first needs to bind ATP. If only one protein in the vast protein library were able to bind ATP, then searching blindly for it would be futile. Its discovery would require a miracle. To find out how rare ATP binding proteins are in the library, Keefe and Szostak used a chemical technology that can create many different proteins, each one with a different and completely random amino acid string, a process equivalent to pulling random volumes from the shelves of the protein library. The random proteins these researchers generated were all eighty amino acids long. Because there are more than 10^{104} such proteins, no experiment could create all of them, but this one created an impressive number, about 6 trillion, or 6×10^{12} random proteins.

Keefe and Szostak found that four of them—unrelated to one another—can bind ATP. Four new ATP-binding proteins out of six trillion doesn't sound like too many, but when the proportions are extrapolated to the number of potential candidates, the number is much larger. It comes out to more than 10^{93} proteins—a 1 with 93 zeroes—that can bind ATP. The problem of binding ATP has astronomically many solutions.[32]

John Reidhaar-Olson and Robert Sauer from the Massachusetts Institute of Technology approached the same problem from a different tack. They focused on a regulator protein that can shut down genes in a virus that infects bacteria. The DNA of this virus—bacteriophage lambda—encodes proteins that help it replicate and kill its host bacterium. But this virus can also remain dormant inside the bacterium, using this off switch to shut down its genes until the time is ripe to divide and kill the host. This time usually comes when the host falls on hard times— starved, poisoned by antibiotics, or irradiated with too much ultraviolet light. The virus then starts to replicate, and its children abandon the cell, rats scurrying from the proverbial sinking ship.[33]

Reidhaar-Olson and Sauer explored a neighborhood of the protein library near this viral off switch, creating many random amino acid sequences in this neighborhood, and asked which of them produced a switch that works, one that can shut down the viral genes. From this information, they calculated that more than 10^{50} texts in the library encode a working off switch. When they tried a similar approach on a different protein, an enzyme needed to synthesize amino acids, they found that some 10^{96} amino acid strings can do this enzyme's job.[34]

Nature's antifreeze proteins gave us a hint, and laboratory experiments like these prove it: Problems like that of binding ATP, shutting down a virus, or catalyzing a chemical reaction don't have just one solution. Or even a million solutions. They have astronomically many solutions, each embodied by a different volume in the protein library.[35] To imagine the sheer number of these solutions is difficult, but that says more about the limits of our imagination than about life's innovability.

It's not enough to know, of course, that a library contains a virtually inexhaustible supply of books describing solutions to a particular problem. We also need to find out where these solutions are and how they are organized—in meticulous stacks or thrown together in unruly piles. And for that we need to move beyond laboratory experiments. Even though such experiments can create and test impressive numbers of different proteins, these numbers vanish into insignificance compared with those found in nature, which churns out new proteins every day in countless trillions of live organisms. Every one of these organisms harbors thousands of proteins, and each is only the last link in an unbroken chain of protein creation that goes back billions of years.

Protein scientists have been aware of this bounty for many years. And they throw themselves at it with gusto, like kids in a warehouse-sized candy store. And what these scientists have learned about protein creation in thousands of different organisms goes far beyond laboratory experiment. The oxygen-ferrying hemoglobin we already encountered in the bar-headed goose illustrates how far.

The humble function of hemoglobin, binding and releasing oxygen as it shuttles between the lung and the body's tissues, belies its near-universal importance. Hemoglobin is a member of a large family of oxygen-binding proteins, the *globins,* which are vital not only to us but also to many other mammals, birds, reptiles, and fish.[36] Countless generations have elapsed, parents followed by their children, grandchildren, and innumerable generations of great-grandchildren, since all these organisms shared a common ancestor. During this succession of generations, the DNA that encodes hemoglobin and all other proteins has replicated countless times. The copying errors it suffers each generation are rare—about one for every forty million copied DNA letters in our cells[37]—but given enough time, all genes in a genome will suffer errors that alter the proteins they encode.

An altered amino acid text that prevents a globin from folding also prevents oxygen from traveling where it is needed. In other words, it spells death. But an altered protein does not always lose its function and meaning. Some alterations impair neither folding nor function, and get passed on to the next generation.[38] Over thousands and millions of generations, copy error after tolerable copy error can thus accumulate and slowly change a protein's amino acid sequence..

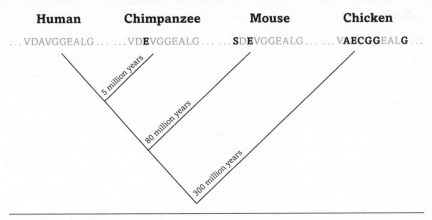

FIGURE 11. Proteins change in time

Figure 11 shows a snippet of ten amino acids from human hemoglobin and from three of our animal relatives.[39] Each letter in the figure is taken from the twenty-letter alphabet that scientists use to abbreviate amino acids, V for valine, A for alanine, and so on. We and our closest relatives, the chimpanzees, shared a common ancestor some five million years, or roughly 200,000 human generations, ago—a huge amount of time compared to a human life span, but precious little in evolution.[40] And because little time means few errors, the globin text of chimpanzees has not changed much since then. In the figure's snippet, the only difference is that chimp globins harbor the amino acid glutamate (letter E, in black) in place of the human alanine (A).

The human lineage split from the mouse lineage some eighty million years ago. Mouse globins thus had more time to accumulate changes than chimps', which shows in the two amino acid differences of figure 11 between mice and humans. The chicken lineage separated from us even longer ago—almost three hundred million years—and accrued six altered amino acids.[41]

Millions of other organisms harbor globins, not only warm-blooded vertebrates but reptiles, frogs, fish, sea stars, mollusks, flies, worms, and even plants. Some of these organisms grow on the same twig of life's gigantic tree and have a recent common ancestor. Their globin texts shared most of their journey through time, split only recently, and are still similar. Others lie on different branches, share more distant ancestors, and harbor globins with different texts.[42] But however different these texts are, each of them works just fine, since otherwise it would not have survived. Each surviving text encodes a different solution to the problem of binding oxygen.[43] And every millennium that life continues, it travels further and further through the library of proteins, discovering ever-new globin texts in its blindly groping evolutionary journey.[44]

To see how far this journey has already led the globins, consider some of our most distant relatives: plants, some of which indeed have globins, even though they have no blood.[45]

Legumes like soybeans, peas, and alfalfa can extract vital nitrogen from its nearly unlimited supply in the air. (Most other plants need to extract nitrogen from the soil, where it is often scarce, unless a farmer has applied fertilizer.) For this purpose, legumes employ bacteria that live in clumps of tissues around their roots, and that harbor a special enzyme that converts airborne nitrogen gas into ammonium, the same ammonium that nitrogen fertilizers contain. This ingenious symbiosis has only one problem: Atmospheric oxygen destroys the enzyme. To protect those enzymes, plants manufacture globins, which keep oxygen safely away from the bacteria.

Plants and animals dwell on different major branches of life's tree, because their common ancestor lived more than a billion years ago. Their globins are staggeringly different, which reflects their long and separate evolutionary journey. For instance, the globins from lupins and insects differ in almost 90 percent of their amino acids. Yet these globins not only bind oxygen, they also still fold into the very similar shapes of figure 12. The fold on the left is from a legume, the one on the right from a midge, a tiny two-winged fly. Both proteins have several spiral-staircase-like helices that are arranged similarly, such as the two helices that run in parallel

Plant   **Insect**

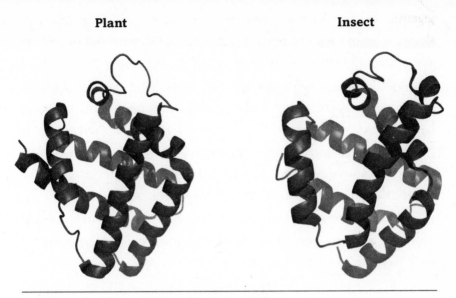

FIGURE 12. Two hemoglobins with similar folds

from the upper left to the lower right. The image does not do full justice to just how similar these globins are. If you rotated these molecules to place one exactly above the other, their atoms would occupy almost exactly the same places. Despite more than a billion years of separation, these globins still fold the same way.

The amino acid differences between these globins are extreme, but not unusual. Even globins from some animals, for example those of clams and whales, can differ in more than 80 percent of their amino acids.[46] Despite these differences, though, these and thousands more globins from other organisms are connected by a network of unbroken paths through the protein library, paths that began at their common ancestor, took one amino-acid-changing step at a time, but left the text's meaning unchanged. You will recognize a theme that we already encountered in the metabolic library, where evolution could travel far and wide without losing the meaning of a metabolic phenotype. The steps evolution takes through the protein library are different—single amino acid changes instead of horizontal gene transfer—but the principle is the same. A *genotype network* connects globins and extends its tendrils far through the protein library. Evolution can explore the library along this network without falling into the deadly quicksand of molecular nonsense.

When it comes to forming vast and far-reaching genotype networks, globins are not an exception but the rule. Enzymes with the same fold, catalyzing the same reaction, and sharing the same ancestor *typically* share less than 20 percent of their amino acids. We know this because scientists have mapped the location of texts encoding thousands of known enzymes in the library. By cataloging these texts, we can map the paths of genotype networks in the library, which reveals that some networks can reach even further through the library than globins, and none more so than that of TIM barrel proteins. Their name is an acronym for triose phosphate isomerase, an enzyme that helps extract energy from glucose, and their fold is called a barrel because its sheets and helices are arranged like the staves of a barrel. The stunning fact is that some enzymes with this fold do not have a single amino acid in common. They occupy opposite corners of the protein library—texts that do not share a single letter—yet carry the same chemical message.[47] Proteins like these are a bit like innumerable versions of *Hamlet,* all of them equally stageable, while sharing only a few hundred—or even none—of the play's four thousand lines.

Thousands of proteins from nature's laboratory tell a similar same story: When a problem can be solved with new proteins, be they enzymes, regulators, or transporters like hemoglobin, the number of solutions is too large to count. And all of those proteins are connected by a vast network of amino acid texts that spread throughout the protein library. We know thousands of proteins from some of these networks, but they are grains of sand on a vast beach of the unknown—most of the many trillions of proteins that share the same phenotype. Some of these unknown proteins belong to long-extinct organisms. But most have not even formed yet. The four billion years of life were not nearly enough time—they would suffice to create only some 10^{50} proteins, a vanishingly small fraction of all texts in the protein library.[48] Life's enormous tree and all its proteins, however vast and beautiful, is but a smeared reflection in a filthy mirror, a faint shadow of the vast Platonic realm that genotype networks inhabit.

In chapter 3, we saw that genotype networks help evolution's billions of readers explore different and far-flung neighborhoods of the metabolic library. Through these networks, some of the library's explorers find innovative texts with new phenotypes, even though others get knocked off a network and die. Genotype networks might do the same for proteins, but only if the neighborhoods of the protein library are diverse.[49] Otherwise, an evolving population of proteins might as well stay wherever it is. No need to explore the library if its different stacks host the same books.

Do the shelves near each protein in the library contain texts with similar meaning, a bit like modern suburbs with their identical cookie-cutter homes? Or is each neighborhood more like a medieval village, with unique buildings and their individual charm, containing proteins with unique new functions? Until recently, we had no idea, even though decades of protein research allow us to answer this question with computers that can mine mountains of protein data.

Answering this question needs more than just computers. It also needs a librarian's love of texts. A young Chilean researcher named Evandro Ferrada brought just this love to Zürich when he joined our group of researchers to get his Ph.D. He had already studied proteins and become skillful at mining huge protein databases for information about proteins, from their folds to their smallest atomic details. I had seen Evandro's quiet, pensive personality before, in people whose minds constantly grapple with the deep mysteries of life. Perhaps this is why he agreed to work on this problem, because the structure of protein space is just such a mystery: one that not only is challenging and profound but also can be unraveled. What's more, it also holds the secret to protein innovability.

Evandro focused on enzymes because they are an extremely diverse group of proteins—no surprise, since they catalyze more than five thousand different chemical reactions. They are also especially well studied: Thousands of them scattered throughout the library have been mapped. Their locations are precisely known, and we can use computers to analyze them. Evandro asked his computer to choose a pair of proteins with the same fold, but in different places on the same genotype network.[50] He then explored a small neighborhood around the first protein, and listed all known proteins in it, together with their function. After that, he explored the neighborhood of the second protein, and listed all known proteins and their functions in *its* neighborhood. Finally, he compared these lists, asking simply whether they were different, whether proteins in the two neighborhoods had different functions. He then chose another protein pair, yet another pair, and so on, asking the same question for them, until he had explored hundreds of pairs and their neighborhoods.

The final answer was simple. The neighborhoods of two proteins contain mostly different functions, even if the two proteins are close together in the library. For instance, even proteins that differ in fewer than 20 percent of their amino acids have neighborhoods whose proteins differ in most of their functions. The protein library has neighborhoods that are highly diverse, just like the metabolic library. And just as with metabolism, this diversity makes vast genotype networks ideal for exploring the library, helping populations to discover texts with new meaning while preserving old and useful meaning.

Both metabolic and protein libraries are full of genotype networks composed of synonymous texts that reach far through a vast multidimensional hypercube, and both harbor unimaginably many diverse neighborhoods. They have much in common with each other, but little with human libraries. And that's not surprising: They were here long before us.

At least three billion years before us. That's when proteins took over most of life's jobs from RNA. They did so for a good reason. Because they have many more building blocks—twenty different amino acids compared to the four nucleotides of RNA—nature could write more texts with proteins. In an alphabet of four letters, you can write about one million different ten-letter strings, whereas an alphabet of twenty letters allows more than ten trillion such strings—ten million times more. This vastly larger number of protein texts increases further with longer texts. More texts mean more shapes, more chemical reactions you can catalyze, more tasks you can perform.[51]

But RNA *did* come before proteins, and for this reason alone it deserves an honorable place in the pantheon of biological innovation. Without innovations made by the first replicators, we would not be here. And our job would be incomplete without understanding their innovability.

Fortunately, there are many parallels between RNA and proteins that can help us understand RNA innovability. We can organize RNA texts into a hypercubic library—not quite as large as that of proteins, but still formidable—where similar texts are near each other and dissimilar texts are far apart. This library also exists in many dimensions, meaning that its neighborhoods are much larger than in three-dimensional space—near any one text are many others. The meaning of many RNA texts is also expressed in a language of shapes, because RNA chains are highly flexible, like proteins. They bend and twist in space, organizing themselves into elaborate folds, like proteins.

Unfortunately, the parallels end with the recalcitrance of RNA molecules to reveal their shape. Experiments have traced this shape only for a few hundred RNAs, a paltry number compared to the many thousands of proteins whose form and function we know. Therefore, what we can do for proteins—compare many naturally occurring molecules to map the library—is not yet possible for RNA.[52]

Thanks to the Austrian scientist Peter Schuster and his associates, though, the RNA library is not a lost cause. One of the grandfathers of computational biology in Europe, Schuster is now a retired professor at the University of Vienna, where he taught since the 1970s. A first encounter with Schuster seems to confirm the stereotypes that many Europeans have of Austrians. A jovial man with a generous girth and a wry sense of humor, Schuster would not have been out of place in the traditional Viennese cafés of the last days of the Austro-Hungarian Empire, where formidably well-educated polymaths held forth on everything from psychoanalysis to quantum theory. He is a scientist in that tradition, a purebred intellectual conversant with a broad variety of subjects. Not taking himself too seriously, Schuster opines with a tongue-in-cheek attitude, peppering the gravest discourse with humorous asides. He epitomizes an oft-repeated saying about how Austrians view life and its many challenges: "The situation may be hopeless, but it is never serious."

There's a broad mind and an incisive intellect, however, beneath the surface of Schuster's jovial demeanor. He was among the first to propose how an RNA world might have originated.[53] And his research group developed computer programs that predict an important aspect of an RNA text's molecular meaning, its secondary structure phenotype.[54]

RNA secondary structure is what emerges first when an RNA string folds. As the string twists and bends and curls, some of its nucleotides pair with one another and create short stretches of double helices in the molecule, much like DNA's famous spiral staircase. The secondary structure is a pattern of multiple such helices connected by stretches of intervening single-stranded text, all formed by a single molecule. Like the sheets and helices of proteins, these helices are the flowers that self-organize into the final bouquet of a three-dimensional fold.[55]

Not only was Schuster able to compute RNA's secondary structures from their nucleotide sequences, but his group's computer programs were also blazingly fast. They could predict hundreds of these molecular shapes within seconds. (To this day, we cannot do this for the more complex three-dimensional RNA fold.) With programs as fast as this, one can begin to map the RNA library. And even though we are still miles from understanding RNA's complete fold and function, the secondary structure is very important on its own: If a mutation in the letter sequence of an RNA molecule disrupts its secondary structure, the molecule can no longer fold properly in three dimensions. Secondary structure is *essential* for the molecular meaning of RNA molecules, just as there can be no bouquet without flowers. And that's a very good reason to study it.

Schuster's researchers found a bewildering number of potential molecular meanings in the RNA library, all of them expressed as shapes. For example, RNA strings that are merely one hundred letters long can already form 10^{23} different shapes. Many natural RNA molecules are much longer, and such longer texts can form many more shapes.[56] What is more, texts with the same shapes are organized much like in the protein library. They form connected networks that reach far through the library, allowing you to revise any one text in little steps, radically, while leaving its molecular meaning unchanged.[57] And just as in the protein library, different neighborhoods are more like medieval villages than cookie-cutter suburbs. Each neighborhood contains many different shapes, and any two neighborhoods do not share many of them.[58] All this hints that innovability in RNA follows the same rules as in proteins. And recent experiments show that this is indeed the case.

In an ingenious experiment performed in the year 2000, Erik Schultes and David Bartel from the Massachusetts Institute of Technology blazed a trail through the RNA library.[59] The experiment started from two short RNA texts with fewer than a hundred letters each. The texts are far apart in the library and differ in many letters, but they are not just any two strings. Both molecules are enzymes—*ribozymes,* because they are composed of RNA rather than protein. Each of them wiggles into a different three-dimensional shape and catalyzes a different reaction. The first molecule can cleave an RNA string into two pieces, while the second does the exact opposite, joining two RNA strings by fusing their ends with atomic bonds. Let's call these enzymes the "splitter" and the "fuser."

If you already had a splitter, and you needed to find a fuser somewhere in the library, would that be easy or hard? And what about the opposite, creating a splitter from a fuser? In other words, can you create a specific molecular innovation from either one of these molecules by exploring the library as evolution would? If you were ignorant about genotype networks, you would think that should be impossible, because the two molecules are far apart. And even if were possible, it might be exceedingly difficult, since a single misstep that creates a defective molecule spells death in evolution.

Undaunted, Schultes and Bartel started from one of the molecules and walked toward the other, modifying its letter sequence step by step while requiring that each such step preserve the molecule's function, just as natural selection would demand. They used their chemical knowledge to predict viable steps through the library, manufactured each candidate mutant as an RNA string, and asked whether it could still catalyze the same reaction as its ancestor. If not, they tried a different step.[60]

What they found may no longer surprise you. Starting from the fuser, they were able to change forty letters in small steps toward the splitter without changing the molecule's ability to fuse two RNA strings.

And starting from the splitter, they could also change about forty letters in small steps toward the fuser without changing *its* ability to split two RNA molecules. About halfway between the two molecules, something fascinating happened: Fewer than three further steps completely transformed the function of either molecule. They changed the fuser into a splitter and vice versa.[61]

Like many good experiments, this one carries more than one powerful message. The first is that many RNA texts can express the molecular meaning of the starting fuser and splitter molecules. Second, trails connect these molecules in the library, and they allow you to find a new meaningful text, even if each step must preserve the old meaning. (Genotype networks make all this possible.) Third, while you walk along one of these trails, the innovation you are searching for will appear at some point in a small neighborhood near you.

The experiment used a single reader to explore the library, not the huge populations that do so in real evolutionary time. And what's more, this reader did not take blind, random steps, but was guided by the biochemical knowledge of expert scientists—its steps were designed to stay on the genotype network. This left a lingering doubt in my mind. Could genotype networks also help real evolution—blindly evolving RNA populations—innovate? It would take another ten years to find the answer, which came from an evolution experiment in my Zürich laboratory.

Most people think of evolution as glacially slow, unfolding on time scales much longer than our brief life span. While that is true for human evolution, where a mere fifty generations span a thousand years, many other organisms have much shorter generation times, such as *E. coli*, which reproduces every twenty minutes. Fifty of its generations pass in less than a day. And an RNA molecule can replicate in mere seconds, using the kind of molecular copying machine that replicates DNA.[62] You could fit thousands of its generations in a single day.

Fast-replicating organisms and molecules allow ambitious experiments to reenact evolution in the laboratory. Such *laboratory evolution experiments* monitor how evolution transforms entire populations over many generations. RNA molecules are especially attractive for such experiments, and for the same reason that they were central to early life. They contain both a genotype that can replicate and mutate, and a selectable molecular phenotype, in a single, extremely compact, evolvable package.

In the fall of 2008, I interviewed candidates to conduct a laboratory evolution experiment that could put my nagging doubt to rest. One of them was a young American scientist who stood out, not just because of his credentials but also because he showed up for his interview in hiking boots. While most academic scientists dress informally and would disdain the rigid dress code of other professions, that was still a bit unusual. At least it spoke of a healthy self-confidence.

The young man, Eric Hayden, had just finished his Ph.D. at the University of Oregon, where he had done excellent work on RNA enzymes. A chemist with little prior knowledge about evolutionary biology, he exuded a deep curiosity about innovation. He was instantly likable, thanks to an open face and a smile that lit up the room. After a brief conversation I asked him to chat with other researchers in my group to see whether he would be comfortable in our community. He must have felt right at home, because an hour later he returned to my office in socks—the boots were too warm, he explained.

Eric got the job, and I have never regretted my decision. He was deeply knowledgeable about RNA, a careful experimenter, and a supremely pleasant human being. I felt privileged to work with him.

In my research group, Eric studied a ribozyme, an RNA enzyme that helps some bacteria express their genes. This enzyme recognizes RNA strings with a specific letter sequence, cleaves them, and then attaches one of the string's fragments to itself.[63] (Many organisms harbor molecules that recognize and cleave specific DNA and RNA texts, for purposes as varied as destroying the foreign DNA of infecting viruses and joining shorter snippets of DNA to form larger meaningful texts.)[64] I had a simple question for this enzyme: Could genotype networks help transform it to recognize a new RNA molecule?

To find out, Eric created more than a billion copies of this enzyme—all of them would fit comfortably into a teaspoon—and used a molecular copy machine to replicate each molecule in this population. This machine is imperfect, because it occasionally makes copying errors, thus sprinkling the population's ribozymes with mutations. Eric then used a chemical trick to let only some of those mutants replicate—those that could still react with their RNA target. The trick satisfied natural selection's key requirement that a molecule's function be preserved for its survival to the next generation.

Eric's experiment cycled multiple times—think of each cycle as a generation—between this error-prone replication and selection. Before the first cycle, all the molecules in his population were identical, like a billion readers hunched over the same volume in a library. After the first generation, many molecules had already mutated, and only some mutants had survived. The survivors mutated further in the second generation, and so on, until after a mere ten generations, the molecules in the population differed from the starting RNA by an average of five letters, and some by as many as ten letters. The billion readers had spread out in the library.

This simple observation concluded a first part of Eric's experiment. The population now contained many different RNA molecules, all connected to their parents through a series of single-letter changes. Forced by selection, the molecules in the population had preserved their phenotype, even though their genotype had changed. Because the population had spread through the library, Eric's experiment had shown that a genotype network exists for RNA enzymes with this phenotype.

The second part of the experiment involved two populations. The first population was the one I just described, while the second was like the starting population, with all identical molecules. Eric supplied both populations with a new and different RNA string to cleave. In this string he had replaced one atom of phosphorus with an atom of sulfur, which makes the enzyme's job much more difficult. He then evolved both populations separately, through multiple rounds of replication and selection, except that now he selected only molecules that could cleave the *new* RNA string as well. He then asked which populations got the hang of this new task more quickly, the first, spread out over a genotype network, or the second, concentrated in one spot.

If genotype networks help innovation, then the first population should do better, because its members can explore more neighborhoods in the RNA library. And this is exactly what Eric found. The spread-out population found an RNA molecule that excelled at the new task eight times more rapidly than the highly concentrated population.[65]

Eric's experiment also harbored another surprise. This one occurred when we read the letter sequence of the best new molecule that evolution had found, the molecule that was superior to all others at the new job.

Many researchers have studied the RNA enzyme with which we started our experiment. It is a short molecule with only some two hundred letters. We know its letter sequence, its fold, what it does, and how it does it. We know most of what you might want to know about it. And we precisely controlled the environment in which the enzyme evolved, down to the concentration of every molecule, for many generations. With this near-perfect knowledge, you might think that we should be able to *predict* how the molecule would change to solve its new task. If you knew a machine inside out, every single cog, bolt, lever, and spring, and how they work together, you would surely know how best to improve the machine.

We didn't. The solution that nature came up with to improve our enzyme was completely unexpected. To this date, we do not fully understand why it works best.

Such surprises occur time and again in laboratory evolution experiments. No matter how well a molecule is studied, no matter how simple the experiment, no matter how precisely controlled, nature never ceases to surprise. Even simple enzymes are more difficult to understand than most human machines.

But although we failed to predict a specific solution, we had managed to predict something more important: that a genotype network could accelerate the population's discovery of this solution. And this prediction was right on target. We can predict innovability even if we cannot predict individual innovations.

Many nonscientists feel troubled when science demystifies nature, identifies the laws that rule it, and thus takes away from their feelings of wonder and awe about the world. In the words of the poet John Keats, a scientist is the sort of killjoy who would "clip an Angel's wings" and "unweave a rainbow."[66] That sentiment surely was one reason why Darwin's theory met resistance, but an experiment like this shows us that we can have it both ways.[67] Science can explain *general* principles of innovability even if it cannot predict any *individual* innovation. Understanding innovability can leave the magic of innovation intact. And that, by itself, is reason for wonder and awe.

CHAPTER FIVE

Command and Control

I t's hard to beat milk as a metaphor for goodness. Lady Macbeth's husband is too full of the "milk of human kindness" to commit regicide, the third chapter of Exodus promises the Hebrews a "land flowing with milk and honey," and to this day we call harmless things "safe as mother's milk." But for more than half of the world's population, a healthy glass of milk is decidedly not good. It means bloating, gas, and diarrhea. The reason is a lack of lactase, the enzyme that prechews the milk-sweetening sugar lactose for our bodies. Without it, bodies cannot break down lactose, leaving it to gut bacteria that happily scavenge this unused fuel and leave waste products with nasty side effects.

When those lactose-intolerant adults were babies, they could digest the sugar in their mother's milk just fine. Their lactase gene was turned on—in technical terms, it was *expressed*—which means that their bodies transcribed the DNA instructions for lactase into RNA and translated this RNA into the needed enzyme. The bodies of lactose-intolerant adults have switched the lactase gene off permanently and no longer express it. Genes like the lactase gene, which our bodies can turn on or off, are *regulated* genes.

For most of human history, the "off" position in adults was the norm. If you are lucky enough to tolerate lactose, you have a mutation in the lactase control region, a stretch of DNA near the lactase gene that leaves the lactase gene turned on well into adulthood. Chances are that your distant ancestors were milk-drinking cattle farmers, because mutations that cause lactose tolerance first spread through pastoral populations, like those of East Africa and Scandinavia. And they spread blazingly fast, from zero to more than 90 percent of some populations, in a blip of time, the eight thousand or so years since humans first discovered the pastoral lifestyle. They are among the strongest recent signatures that natural selection has left in our genomes.[1]

Surprising as it may seem, lactose-induced indigestion is deeply connected with innovation. What connects them is regulation—the tuning of the activity of molecules like the lactase gene. Accounting for much more than intestinal upset, regulation is also complicit in the endlessly varying forms of organisms, the gracefully undulating umbrella of a jellyfish, the lethal torpedo of a shark's body, the slender stem of a rose, the gargantuan trunk of a redwood tree, the deadly coil of a viper, the light-footed legs of a rabbit, and the soaring wings of a bird. Regulation has come a long way from its murky origins in the first cells, where it balanced the growth of a membrane container with that of an RNA genome. More than three billion years later, regulation is shaping the bodies of every living thing on the planet. And no understanding of innovability would be complete without grasping how new regulation appears.

Although regulation controls the form and function of even the most complex organism, like so much else it is most easily studied in the simplest cells, those of bacteria. This is how two French geneticists, François Jacob and Jacques Monod, won their Nobel Prize. Starting in the 1950s, at a time when the double helix had just been discovered, they showed how primitive bacteria like *E. coli* regulate the expression of the genes that permit them to digest lactose.[2]

Gene expression begins with the kind of molecular copying machine that we briefly encountered in Eric Hayden's experiment in chapter 4. It is a *polymerase* enzyme that makes—the name says it all—a polymer, a stringlike molecule consisting of many smaller building blocks, the four nucleotides that we find in the faithful RNA transcript of a gene.[3] When this polymerase transcribes a gene, it first attaches to the gene's DNA, slides along this DNA letter by letter, and strings together an RNA molecule whose letter sequence is identical to that of the gene.[4] This is also how bacteria express the gene for *their* variant of lactase, an enzyme called beta-galactosidase.[5] (The name is cumbersome, hence it is often abbreviated as beta-gal.) This enzyme cleaves lactose into the two simpler sugars glucose and galactose, from which other metabolic enzymes can extract energy and carbon.

To regulate the beta-gal gene, cells manipulate its transcription with a *transcriptional regulator*. This protein does mostly one thing: It latches on to short stretches of DNA near a gene. Inside the liquid environment of a cell, multiple kinds of regulators drift this way and that, and whenever any one of them encounters a specific DNA sequence—a DNA "word"—it will bind and stick to it. Different regulators have different keywords—the beta-gal regulator recognizes one that contains the letters GAATTGTGAGC.[6]

What enables this recognition is the same folded protein shape that makes enzymes work. Regulator and DNA need to have complementary shapes, a bit like Lego blocks where several small studs on one block fit snugly into indentations on another. The analogy is apt but also limited, because shape is not all that matters. For instance, the two molecules also need to have complementary charges or they may repel each other. And where the standard set of Lego blocks has only a few dozen shapes, molecules have many more, tens of thousands in proteins and even more in DNA, where there are as many shapes as there are possible words.[7]

What is more, unlike Lego blocks, many molecules spontaneously

No lactose

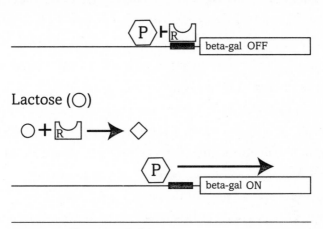

Lactose (○)

FIGURE 13. Gene regulation

change shape, not only when they vibrate like enzymes but also when they bind one another. This shape change is similar to what happens when you insert the right key in a lock: Only then can the lock's cylinder turn and open a door—although in molecules nothing but heat is doing the turning.

The regulator's Lego-like binding modulates beta-gal production in the simplest possible way: It creates a roadblock for the polymerase, because the regulator's keyword is placed right where the polymerase starts to transcribe (upper part of figure 13).[8] When no lactose is around, the regulator (R) binds this word and blocks the polymerase (P) from reading the gene— the gene remains turned off.

To use lactose, cells need to get rid of this roadblock whenever the energy-rich substance appears. To understand how they do that, it helps to know that the regulator can bind not only DNA, but also another molecule—like one Lego block that can connect to several others. The other molecule is lactose itself. And when the key of lactose binds the lock of the regulator, the regulator changes shape (see the diamond shape in figure 13) and is no longer complementary to DNA (lower part of figure 13). It detaches from the DNA, which the polymerase can now transcribe freely, letter by letter, into the RNA from which the cell will manufacture the beta-gal protein. In sum, the beta-gal gene is on—beta-gal is produced—whenever lactose is available, but otherwise the gene is off, because its transcription is blocked.

Beta-gal is great stuff. But it isn't cheap. A cell that expresses beta-gal doesn't just contain a few dozen of the beta-gal proteins, but rather some three thousand identical molecules, each of them comprising more than a thousand amino acids that need to be manufactured and strung together, which has to be paid for in molecular materials and energy.[9] Common sense suggests that cells should regulate beta-gal to avoid wasting these materials, but if common sense were the surest guide to nature's ways, biologists would have little to do. The cost to produce beta-gal constantly may well be negligible, given the millions of other molecules that cells manufacture. And leaving the gene on all the time could have a real advantage too, a head start when lactose becomes available.

In 2005, Erez Dekel and Uri Alon from the Weizmann Institute in Israel wanted to find the true cost of expressing beta-gal.[10] They tricked cells into believing that lactose was around when in fact it was not. The cells turned on the beta-gal gene without reason, and if that wastefulness makes a difference, they should divide more slowly. And indeed they did, by a few percent. It's a bit as if a cash-strapped developer built a house with an unnecessary swimming pool, which cost him money and materials that should have been used to build other, essential rooms. A better builder would finish faster, sell the house, and start a new one, while the other one is still agonizing over the tiles to choose for the pool.

A construction delay amounting to only a few percent may not seem like much, a single minute or so, added to the twenty minutes that *E. coli* cells need to divide. But that minute will eventually be deadly. A population that initially contains 50 percent of such wasteful cells would contain less than 1 percent after eighty days, and fewer than one in a million after merely three hundred days. Rapidly, inevitably, and fatally, they get washed out by the faster-reproducing cells. This is natural selection in rapid and brutal action.

If regulation matters because it avoids waste, then it should be everywhere. And indeed it is. Think of a metabolism with its hundreds of reactions—lactase catalyzing only one of them—as a sophisticated interconnected network of pipelines. Into this network flow nutrients, out of it flow biomass molecules. Each pipe has a dedicated pump, an enzyme that propels materials through it. A cell can regulate each pump according to its needs. If new nutrients turn up in a patch of soil—a fallen apple, a rotting carcass—the soil bacteria turn up the pumps through which these molecules flow. Once the nutrients are gobbled up, they shut these pumps down. And if more of some nutrients and less of others become available, cells can fine-tune the pumps to the right speed.

The beta-gal gene is *repressed* by a regulator, but other genes are regulated in the opposite way: Cells leave them off by default and *activate* them only when needed—through transcriptional regulators that help rather than prevent DNA polymerase from transcribing a gene.[11] And even though the regulation of transcription is the most important kind of regulation, there are many others. Cells regulate how fast they manufacture proteins from transcribed RNA, how active these proteins are, how long-lived, and on and on. This is perhaps the most convincing evidence for regulation's importance: Life has invented a dozen different ways of doing it.

Imagine the kitchen of a high-end restaurant. The pantry is well stocked with all manner of vegetables, meats, fruit, fish, cooking oils, spices, and flavorings. There are enough ingredients to create every conceivable dish, from the mundane to the exotic, each one nutritious and delicious. The executive chef wants a complete supply of everything, all the time. One of regulation's roles in metabolism is that of a penny-pinching manager, requisitioning just the right amount of ingredients, anxious not to squander money on even as little as an extra potato.

But regulation is much more than that. It also provides the recipes, the kind that call for cooking a cup of beans with two cups of chicken stock and a dash of salt in a 350-degree oven for thirty minutes. Each recipe is a sophisticated gene expression program encoded in the genome. It tells cells just when and how much to make of each protein ingredient in an organism.

If comparing a blue whale to a soufflé seems offensive, consider the complexity of life's recipes. Each cell type needs many more ingredients than even the most complex dish, thousands of them, in amounts so finely tuned and so exquisitely timed that not even the most skilled five-star cook could hope to follow life's recipes. What is more, evolution has untiringly created new recipes that brought forth ever-new dishes, innovations in cells, tissues, and organs, as well as entirely new *kinds* of bodies that emerge through the shifting and enormously complex patterns of regulation.

Regulatory recipes are the business of developmental biology, a branch of biology that studies the near-magical process of creating a body from a single cell. It ponders how the cells in a body form not just an amorphous, shapeless blob, but organs like brains, livers, hearts, and lungs in animals, and stems, leaves, roots, and flowers in plants. Each organ has a highly specialized task, and each contains many specialized types of cells. Your heart, for example, includes contracting cells that make the cardiac muscles pump, connective tissue cells that hold them together, and pacemaker cells that coordinate the heartbeat via electric signals, like a drummer beating the rhythm of the oarsmen on a galley. How do these cell types form from a single fertilized egg cell, and why do they form at exactly the right time and place? How does a cell know to become a pacemaker cell rather than a neuron or a liver cell?

The answer lies in regulation, which guides the development of all organisms. The specialized cells in multicellular organisms get their identity from the proteins they manufacture. Each cell contains complete copies of all our genes, but cells differ in which of these genes they express as proteins.[12] Muscle cells express motor proteins, tiny molecular machines that allow them to contract, which is about all muscles do. Cells in your eye make transparent proteins whose purpose is to transmit light and focus it on the light-sensing retina. Cartilage cells express collagen and elastin, proteins that cushion the bones in your joints and prevent them from rubbing against each other.[13]

The relationship between specialized cells and distinctive proteins seems a simple one, but while cells do express specific proteins, they don't do so exclusively. Any one protein, in fact, can be expressed in multiple cell types. The vitreous body of your eyes—the clear gel between the lens and the retina—produces the same collagen as cartilage, the muscle cells in your biceps express the same molecular motors as the heart, and so on.[14] What gives a cell its identity is not one molecule, but a molecular fingerprint, a combination of hundreds of proteins that is unique to a cell. And the identity of an innovative cell type is really a *new* fingerprint, caused by a new pattern of gene regulation.[15]

Because identity-shaping genes can be expressed in multiple cell types, they have not just one but multiple on-off switches, symbolized in figure 14 through several small rectangular boxes. Each of these boxes stands for a different word that can bind its own regulator (open shapes). An example is the gene encoding crystallin, a protein in the fingerprint of your eye lens that also helps your eyes focus (more about this gene in chapter 6). No fewer than five regulators, including a protein called Pax6, bind near it and determine its expression.[16] Some such regulators bind

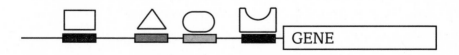

FIGURE 14. One gene, multiple regulators

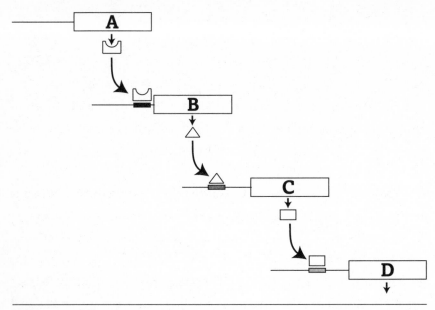

FIGURE 15. A regulation cascade

DNA strongly and also influence transcription strongly, others bind weakly and influence transcription weakly.[17] Like a cabinet of counselors who jockey for influence over a king, these regulators jockey for influence over the polymerase's "decision" to express a gene. Some favor repression, others activation, some are influential, others less so, and the sum of their influences determines gene expression.

And what regulates the regulators? Simple: more regulators. Keep in mind that the regulators of the gene in figure 14 are proteins, and like all proteins, they are encoded by genes that are usually also regulated. The Pax6 regulator of lens crystallin is itself expressed not just in the lens but also in the cornea, the pancreas, and the developing nervous system, where multiple regulators guide *its* expression. And what about these regulators, how are they regulated? Through other regulators. And their regulators? Through yet other regulators. All these regulators often form daisy chains, regulation cascades like that shown in figure 15.

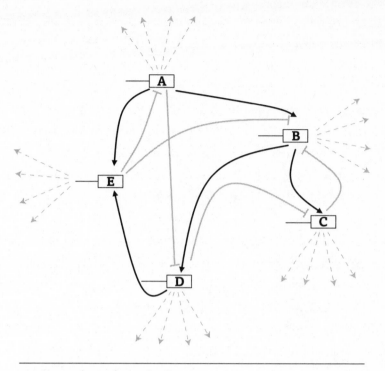

FIGURE 16. A regulator circuit

That would seem enough complexity for one day, but alas, regulation can get *much* more complicated: Regulators form not just linear chains but complex regulator *circuits* where regulators regulate each other. Figure 16 illustrates this idea with a regulator circuit of five genes labeled A through E, each symbolized by a rectangle. To keep it simple, the figure no longer shows where regulators bind DNA, but only which regulators regulate each other, through black arrows indicating that a regulator activates a gene, or through gray lines ending in crossbars for regulators that repress a gene. (As a verbal shorthand, we can say that genes activate or repress one another.) The dashed lines hint at yet another complication: Each regulator can twiddle the knobs of other genes—up to hundreds—outside the circuit. The Pax6 gene is a member of such a circuit, and mutations in it illustrate the power of regulator circuits: Defects in the human Pax6 gene can cause blindness through *aniridia*—a missing iris—as well as an opaque lens and a degenerate retina. The gene has counterparts with similar roles in many animals, including mice, fish, and even fruit flies—despite the great architectural differences between their compound eyes and ours. The fruit fly version of Pax6 is known as *eyeless,* for the simple reason that flies lacking eyeless form no eyes. But even more striking is what happens when flies get too much of a good thing, when biologists turn on eyeless in unusual places while a fly embryo develops: In that case, flies can sprout eyes on their antennae, legs, and even wings.[18]

Figure 16 looks a bit like a wiring diagram an engineer might create. This metaphor, of genes being wired, is useful, even though no real wires run between them. A wiring diagram is a compact way of writing down a circuit's *genotype*, the DNA that encodes its regulators and the keywords that each regulator binds.[19] It allows you to read at a glance that gene A in figure 16 activates B and E, but represses D, B activates C and D, but is repressed by E and C, D represses C, and so on. Inside a cell these influences create a symphony of mutual activation and repression, where each instrument in the orchestra of genes responds to the melodic cues of others with its own notes, until the circuit reaches an equilibrium—like a polyphonic closing chord—where the expression of circuit genes no longer changes and their mutual influences have reached a balance. At this point of balance, some circuit genes are turned off whereas others are turned on. As a hypothetical example, genes A and C in figure 16 might be on, whereas B, D, and E might be off. This state of a circuit (e.g., "on," "off," "on," "off," "off") is the *gene expression pattern* that ultimately creates a cell's molecular fingerprint, because each of the circuit genes regulates many other genes.[20] This pattern is also the circuit's phenotype. It is another one of those phenotypes that can't be perceived directly but needs to be measured with sophisticated instruments. Ultimately, however, it creates the most visible phenotype of all—a body and its shape.

And new gene expression patterns are needed to create new, innovative types of bodies.

Regulatory circuits help build the bodies of organisms as different as the tiny fruit flies that congregate on rotting fruit, the small weedy plant known as the thale cress, and the zebrafish, a striped freshwater fish less than ten centimeters long. None of these organisms is eye-catching, but they have other qualities that make them favorite examples for researching development: They are small and they develop fast, thus allowing us to study many of them in little time.

One lesson they taught us is that regulation circuits can shape a body incredibly rapidly. The larvae of fruit flies, *Drosophila melanogaster*, hatch a mere fifteen hours after a fruit fly has laid its eggs, pupating and metamorphosing into adult flies after as little as seven days. Fifteen hours to produce a complex organism that can feed, crawl, and navigate the world—no wonder that thousands of scientists spend their careers trying to figure out how *that* recipe works.

A fly's body has three major parts that are further subdivided into fourteen segments: a head, a thorax with three segments, and an abdomen with eleven more, where individual segments have special tasks like walking or reproducing. They are not, to most people, especially beautiful or elegant: nothing like the perfection of a bird's wing, or the majesty of a giant redwood. However, those fourteen segments and their special tasks, as distinctive to a fly as a flying buttress is to a Gothic cathedral, or a Doric column to the Parthenon, have produced as much scientific enlightenment as any aspect of the living world. They are still studied by multitudes, from high school biology students to Nobel laureates. That's because they are full of lessons about regulation, many of which apply to all animals.[21]

Before laying an egg, a fruit fly deposits several chemical signals in the egg, tiny snippets of the complicated recipe to form a larva. One of these signals is an RNA copy of a gene called *bicoid*, from which the egg can make a bicoid protein. (Yes, fly biologists use strange names.) Deposited near the egg's front, where the head will later form, bicoid disperses through the egg like a small glob of syrup through a glass of water. Its concentration remains highest at the front and decreases toward the rear.

In addition to bicoid, the mother fly also deposits RNA transcripts of several other genes at the front, and yet other transcripts at the rear, where they spread toward the front. After the mother's job is done, each region of the embryo is bar-coded with a combination of regulators in amounts unique to that region.

After a sperm has fertilized the egg, the egg begins to divide into many cells, and the developing embryo synthesizes proteins from the deposited RNA molecules. In any one cell, the amount of each synthesized protein depends on how much RNA the mother deposited nearby.[22] These proteins are—you guessed it—regulators that can turn other genes on or off, depending on their amount. If a gene's activator, for example, is abundant in the front like bicoid, the gene would be turned on by bicoid in the front, but nowhere else.

Among the many genes these regulators control, some are special, because they encode further regulators, which turn yet other genes on, some of which also express regulators, and so on. What's more, these regulators then regulate each other. They form a complex regulation circuit with more than fifteen different genes.[23] This circuit performs the choreography of mutual regulation I described earlier, creating a pattern of gene expression in which some regulators are expressed and others are not.

An especially remarkable protein in this circuit is called *engrailed*. Guided by the circuit's choreography, it becomes expressed in seven highly regular stripes across the embryo. Seven regions that produce engrailed alternate with seven other regions that don't, and the stripes of engrailed expression demarcate the nascent fourteen segments of the fly. Engrailed and other regulators then control many further genes that specify each segment's identity, specifying whether a segment will carry legs, or support wings, or be part of the abdomen.[24]

All this and more happens in a matter of hours, but it's not only their speedy development that made fruit fly embryos a boon to developmental biology: Until segmentation is well under way, the embryo's cells are not separated by walls. This means that molecules can drift freely through the growing embryo. In most other species, cells are walled off right after fertilization, which makes communication between them more challenging.

Not impossible, though. The male reproductive apparatus in humans—the penis and scrotum—is an instructive example. When a male fetus is about eight weeks old, clusters of so-called Leydig cells release chemical signals called androgens near the area where sex organs will eventually form. These signals are hormones like testosterone that are crucial for sculpting sexual organs. Released from the Leydig cells, they instruct nearby cells to specialize in forming the penis, the scrotum, and, much later, sperm cells. Once an androgen hormone molecule has been released by a Leydig cell, it drifts through the space between cells, and because its chemical structure allows it to penetrate the cellular membrane, it eventually enters another cell. Inside, an androgen receptor, a special protein that can recognize the hormone's shape, is already waiting for it. And when the two connect, the receptor protein changes its shape—yet another turning lock. Its new shape allows the protein to bind specific words on DNA and activate nearby genes. The androgen receptor turns on many genes, some of which make regulators that eventually arouse hundreds of genes to create unique cellular identities within the male sexual organs.[25]

From flies to humans, hundreds of similar signals crisscross every tiny sliver of tissue during every moment of an embryo's development. This unimaginably complex communication process instructs cells about their location and fate, like the bicoid-expressing cells that "know" they are near the front. These signals also command cells to divide, move, swell, shrink, flatten, and, eventually, to acquire an identity and shape a body. And they are involved whenever cells acquire new identities that lead to innovative shapes and body plans.

If we could predict the regulatory dance that shapes embryos from flies to humans, we could predict how organs, tissues, and cells form, and why different organisms have very different body plans. That would be quite a feat. Unfortunately, a circuit's expression pattern can be extremely complicated, even for a simple circuit like that of figure 16. If A activates B and C represses B, whereas B activates C, and D represses C, it's not immediately obvious which gene expression pattern would result. And worse, many circuits contain many more genes than this one, dozens of regulators that form a dense filigree of interwoven regulation, complex beyond our mind's grasp. But there is hope, in the form of computers that can describe a circuit's choreography through mathematical equations, process these equations in their silicon brains, and predict a circuit's gene expression patterns.

One remarkable computational scientist who devotes his life to this task is John Reinitz. I first met John when I was a graduate student at Yale in the 1990s. He was a few years my senior, and what they call a character, chain-smoking filterless cigarettes at a time when smokers were already ostracized, and dressing in ways that made casual Friday look like a formal banquet. He drove a fossilized Volkswagen Beetle, whose rear seats had disappeared beneath a landfill of discarded fast-food packaging. John's nonconformism surely was an asset in his research, because he swam with bold strokes against the mainstream.

At the time, many scientists studied fly embryos, but they used computers for little more than writing their research papers. Instead they studied regulation by changing the DNA text of a gene, or manipulating a regulator in the laboratory, and measuring how such changes altered segmentation. Their experiments were productive: Among other things, they established which among thousands of genes formed the fruit fly's embryo. But their efforts to understand a circuit's entire expression patterns, one gene at a time, were doomed to fail, because a whole circuit is so much more than the sum of its parts. Today this is widely accepted, but in the early 1990s John's efforts to emulate fly development in a computer made him an outsider. His work was ignored by many and belittled by some.

This work was analogous to building one of those flight simulators that are indispensable for training military and commercial pilots—by reproducing not only all the complex machinery of a cockpit, but also disturbances like turbulence and instrument failures. Similarly, John's fly simulator collected mountains of data about the regulators of early fly embryos and how they regulate each other, encapsulated this information in equations, and simulated them in a computer. And like a good flight simulator, this one worked—not a small achievement. It can mimic the early development of fruit flies, and does so at enormously accelerated speed. It can be run over and over again, to tease out patterns that might be missed in isolated examples. And more than reenacting the choreography of a normal embryo, it can also simulate the plane crashes of regulator malfunction, and explain how mutant genes lead to deformed embryos.[26] As I write these lines, John has devoted decades of his life to building this simulator, often in the face of ignorant and condescending peers. His dedication often crosses my mind when I am about to swat a fly. (Then I swat.)

Beyond sharing a backbone and a spinal cord, the more than sixty thousand species of vertebrates, which include fish, mammals, amphibians, reptiles, and birds, have incredibly diverse bodies. This diversity is built on similar structures, however, because all vertebrates can trace their lineage back to a common ancestor that appeared more than five hundred million years ago. For example, the paired fins—one pair in front, one in the back—that help fish push and steer themselves through water gave rise to the arms and legs of animals that crawl, walk, jump, and run on land. And the forearms of some of these animals—dinosaurs— changed into the wings of birds.

Limbs are key innovations of land-living vertebrates. They have three familiar parts, the upper arm and upper leg, the lower arm and lower leg, and the hands and feet. The major bones in our arms and legs correspond to arm and leg bones in horses, dogs, eagles, bats, pigs, crocodiles, and many other animals. Transform their sizes, as evolution did, and many specialized functions become possible, such as the slender limbs of horses custom-made for running, and the light bones of wings perfected for flying.

Limbs—old and new—owe their existence to a family of regulators that are used in building the bodies of thousands of organisms, from jellyfish to humans.[27] Though these regulators are essential for normal body development, the name for the genes that encode them—homeobox, or Hox, genes—comes from their role in homeosis, a process that creates malformed organisms when these genes are mutated, such as flies with (useless) legs sprouting from their heads in place of antennae. For better or for worse, changing life's recipes can have dramatic effects.[28]

The homeobox is a protein sequence of sixty amino acids that binds DNA and allows Hox regulators to control gene expression. In organisms as different as fruit flies and humans, these regulators are lords over hundreds of other genes that give texture to cells, tissues, and organs. Hox regulators also regulate one another's expression. That is, they form a regulator circuit like that of figure 16, but much more complex, because animals can have forty or more of them. This circuit shapes important parts of many bodies—including ours. Among these parts are the

thirty-three vertebrae in our spinal column and their unique identities—
two vertebrae in our neck with flexible joints, twelve vertebrae in our
thorax with attached ribs, and so on.

The Hox gene circuit expresses different combinations of genes in
the neck, thorax, and abdomen as our backbone develops in the womb.
Each of these combinations is a *gene expression code,* an on-off pattern of
Hox genes that is specific to each body region. One on-off pattern specifies neck vertebrae, another specifies thoracic vertebrae, and so on.

Hox genes mold not just the human body but also the bodies of vertebrates like pythons and other snakes whose body plan—another ancient
innovation—allows them to slither, burrow, and swim. Some snakes have
more than three hundred vertebrae, most of them identical and rib-
carrying like our twelve thoracic vertebrae. Hox genes are responsible for
these differences between snakes and other animals: In most vertebrates,
the Hox code for the thorax is expressed in only a small region of the
embryo, but this region stretched like a rubber band when snakes evolved
from lizards more than a hundred million years ago. The thoracic Hox
code became expressed along most of the main body axis, and allowed
snakes to build the hundreds of vertebrae that define their new body
plan.[29]

Hox genes shape not only the main axis of an organism's body—the
axis defined by the backbone in vertebrates—but evolution also co-opted
them to mold another ancient innovation, the fins of fish.[30] And it did not
stop there. Over millions of years, evolution transformed these fins into
limbs, by altering, refining, and differentiating the fins' Hox code. Eventually it created a three-part code in organisms that walk or fly, a specific
combination of Hox genes for the upper arm, another for the lower arm,
and a third for the hand. We know all this from the effects of mutations
that garble this code and appear as horrific birth defects that are well
studied in animals. When two genes called Hoxa11 and Hoxd11 are not
expressed while limbs develop, the results can be no lower arms at all, or

a hand sprouting from somewhere near the elbow. Likewise, missing fingers or palms in a newborn can result from a failure to express two hand-specific Hox genes—Hoxa13 and Hoxd13. If the expression of a third group of Hox genes fails, only the upper arm will form.[31]

Most of the time, though, Hox genes do their job very well. And they do it in an impressive number of locations, helping form structures from the pelvis to the vertebrate brain. They also help to build the bodies of organisms as different as shrimp, jellyfish, worms, and even fruit flies, where the Hox circuit is as important as the segmentation circuit. In fact, the two work sequentially. After the segmentation circuit establishes segment *numbers,* the Hox circuit specifies segment *identities*—which segments will carry legs, which ones wings, and so on. And these circuits are only two among many that flies and most animals use to build their bodies, and have used ever since the first animals emerged hundreds of millions of years ago.

Hox gene circuits were instrumental in the origin of new body *parts*—like limbs—as well as new body *plans,* like those of snakes. How exactly these innovations originated may be forever lost in the mists of life's deep history, but one principle is crystal clear: They originated through changes in regulation.

The same principle is just as clear in other, more ancillary innovations.

Imagine a slender lizard that weaves its way through a dense meadow hunting for its next meal when suddenly a huge pair of eyes stares into its face. It freezes, knowing that it will be torn to pieces in a moment. But then two wings flap, and the eyes are gone like a mirage. No predator was near, just two enormous colored spots on the wings of a tasty butterfly.

The eyespots of butterflies are lifesaving bluffs, formed by an unusually versatile regulator protein called *distalless.*[32] A member of a circuit that molds the legs, wings, and antennae of flies, distalless has also been co-opted to paint eyespots on the butterfly's wings. We know that distalless is part of an eyespot-specific expression code, because developing butterfly larvae make distalless exactly where the eyespots will later form. Some

simple leaf   **dissected leaf**

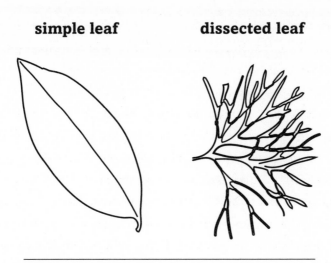

FIGURE 17. Leaf shapes

butterflies have smaller eyespots, others larger eyespots, some have only one eyespot, others several. Regardless, developing butterflies unfailingly express distalless in the location of their eyespots. And distalless is really a cause of eyespots rather than just correlated with their appearance: If one transplants distalless-producing cells of a developing wing to different wing locations, development will paint an eyespot there.[33]

The cathedral of a butterfly's body is built by regulation, from the nave of its main segments to the gargoyles of its eyespots. So are bodies with completely different blueprints, like those of plants, with their roots, stems, flowers, and leaves. When flowering plants first originated more than two hundred million years ago, they had simple leaves whose blades were undivided and formed one continuous surface. Later, simple leaves gave rise to the innovation of a dissected leaf, where many small leaflets subdivide a leaf blade (figure 17).

Dissecting a simple leaf into leaflets offers more than one advantage. Dissected leaves have a greater surface area than simple leaves. They can absorb more carbon dioxide for photosynthesis, which allows plants to grow faster, and they can prevent leaves from overheating in hot environments, which can slow down photosynthesis and damage the leaf.[34] If dissected leaves are so useful, we might expect them to appear more than once in evolution, and indeed they did: Dissected leaves arose more than twenty times in flowering plants alone.[35]

Each time this innovation required changes in regulation. When a plant seedling germinates and pushes through the soil, a tiny speck of tissue at its very tip contains dividing cells that enlarge the seedling and push it upward. It is here that leaves begin to form. Before you can see a nascent leaf with the naked eye, a cluster of multiple cells—the leaf primordium—is already set aside around the tip to become a leaf. Cells in this primordium express a regulator protein called KNOX. When Angela Hay and Miltos Tsiantis from Oxford University manipulated this protein in the modest weed known as the hairy bittercress, which sports dissected leaves, they found how crucial this regulator is. By decreasing the amount of KNOX, they could reduce the number of leaflets down to one, creating a simple leaf. Increasing the amount of KNOX created leaves with more leaflets. Plus, they found that KNOX plays this role not just in the hairy bittercress but in several other plant species with dissected leaves.[36]

These examples and hundreds more illustrate the power of regulation to innovate. The lab notebooks of thousands of researchers and the pages of dozens of scholarly journals are overflowing with research on regulators like KNOX in plants, distalless in butterflies, and engrailed in fruit flies. Our own genome encodes more than two thousand different regulators in dozens of separate circuits.[37] A half century of research has told us how important regulation is to building bodies old and new. It has helped us to understand the natural history of many innovations and the new expression codes behind them.

But a list of examples, however long, cannot go beyond that. Lizards' limbs and fishes' fins are shaped by different variants of Hox circuits—different circuit genotypes—that produce different expression codes. Identifying any one such circuit variant does not explain how evolution found the one whose expression code is best suited for a task. (If there are too many circuit variants, this could be impossibly hard.) What's more, while circuits change little by little in evolution, useful expression codes need to be preserved before new and better ones are found. No list of examples, however long, could tell us how innovation through regulation is even possible.

If the problem is familiar, so is the solution: Study not just one circuit but many, an entire library of circuit genotypes and their expression phenotypes. The texts in this regulation library are the DNA genotypes that encode regulators and the words they recognize. But writing them like that would be unnecessarily long and tedious, as if you described a house through the position of all its molecules, rather than by an architect's blueprint. Much better to write them as wiring diagrams like those of figure 16.

The entire library comprises all possible such circuits—all possible wiring diagrams. To compute its size we need to count these wiring diagrams. That may seem hard, but it is surprisingly easy. Any regulator in a circuit, call it A, can influence another regulator, B, in three principal ways. Regulator A can activate B, it can repress it, or it can have no effect. The same holds for any other pair, say, A and C, or D and E, in the circuit of figure 16. One can activate the other, repress the other, or have no effect on it. These are the only three options. This simple idea takes us almost all the way to counting all five-gene circuits. What's left is to count the number of gene pairs. The circuit of figure 16 has $5 \times 5 = 25$ of them, each with three regulation options.[38] To find the total number of circuits, we then need to multiply three for the first gene pair with three for the second gene pair with three for the third pair, and so on, for all twenty-five pairs. Three multiplied with itself 25 times yields 3^{25}, or more than 800 billion circuits.

An impressive number. Five genes. More than 800 billion circuits. Especially since actual regulation circuits can have many more than five genes. The Hox gene circuits of vertebrates, for example, comprise some forty-odd genes.[39] To count the number of circuits that these genes could form, we use the same idea: Compute the number of gene pairs (40 × 40 = 1,600), and then multiply the number 3 with itself 1,600 times. The magnitude of the resulting number has the ring of familiarity. It is greater than a 1 with 700 zeroes behind it, more zeroes than would fit on this page.

But even that number, impressive as it is, doesn't yet capture all circuits. So far, we assumed that all regulators were equally influential, turning their target genes either on or off. But remember the king's cabinet of counselors. Some regulators can be weak, others strong, and that difference further increases the number of circuits: Any two genes might face not three but five possibilities, no regulation, weak or strong activation, weak or strong repression. In this case, we would need to multiply the number 5—not the number 3—by itself many times. And why stop there? We could distinguish ever-finer gradations of activation or repression leading to ever-increasing numbers of possible circuits.[40] Fortunately, research in my laboratory has shown that these finer gradations of influence don't change the library's organization—a good thing, since even three gradations create enough circuits to fill yet another hyperastronomical library.

The circuit library and its genotype texts have much in common with the metabolic library and the protein library we encountered earlier. Clip or add a wire in a circuit through DNA mutations—remember that these "wires" are not made of metal, but symbolize regulatory connections between two genes that can be altered through DNA mutations—and you create one of the circuit's neighbors, like that on the right of figure 18, where gene B no longer regulates gene D (see the thick black arrow in the left circuit). Each circuit has many such neighbors, more

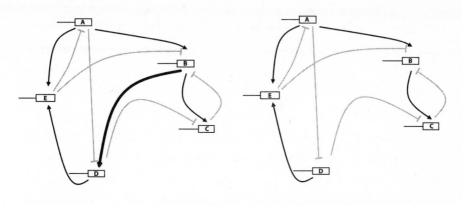

FIGURE 18. Two neighbors in the circuit library

than three thousand for a circuit with forty genes. If we arrange all circuits on the corners of a hyperdimensional cube, one circuit per corner, then stepping away from a circuit is like moving along an edge of this hypercube to the next corner.[41] And many edges lead away from each circuit, because this hypercube also has many dimensions, sixteen hundred for circuits of forty genes. It has even more corners, 10^{700} of them, the number of texts in the entire library of forty-gene circuits.[42]

As in the other two libraries, each circuit on each corner has a neighborhood that includes all texts on nearby shelves—the circuits that differ from it in only one or a few wires. Evolution can easily explore this neighborhood in a few steps, DNA alterations that change as little as one DNA word, and create or destroy regulation between two genes. Walk beyond this neighborhood and you encounter circuits that are ever more distant—another familiar concept. Here, distance is the number of wires by which two circuits differ. Neighboring circuits are closest, and farthest apart are two circuits that do not share a single wire. They are texts in opposite corners of the hypercube library.

Many circuit genotypes will be as meaningless as a random string of English letters. Others may encode meaningful words or sentences, even though the text as a whole may be incoherent, or even destructive like the mutant Hox circuits that create crippled arms without hands. The language of meaningful texts is once again a chemical language, that of gene regulation and expression codes that cells and tissues understand. It ultimately manifests itself in a backbone, a leaf, or a hand, each one a parcel of meaning embodied in flesh.[43] And when evolution creates new embodied meaning, it does so through the kinds of mutations that turn a simple leaf into a dissected leaf.[44]

A circuit's meaning is expressed through the elaborate choreography of gene regulation I described earlier. Starting from a pattern of expressed regulators—like the one a fly imposes on its egg through its chemical signals—circuit genes regulate each other and change this pattern. Genes twinkle on and off until a circuit finds an equilibrium resembling the human sculptures that troupes of circus acrobats build with their bodies. In such a sculpture, the acrobats are in a stationary equipoise where the push of one body equals the pull of another, and where the structure would collapse if only one acrobat were to let go.

After many years of research, we have learned enough about this kind of regulation to compute this equilibrium, as John Reinitz showed with his fly simulator.[45] This means we are ready to read not just one, not a few, but millions of circuits. We can map an entire hyperastronomical library of them.

We already know that this library contains unimaginably many circuits, but the number of their expression codes is not for the fainthearted either. If each gene in a forty-gene circuit can only be on or off, it contributes two possibilities to a gene expression pattern. To calculate the total number of possible expression patterns, we need to multiply two with itself, as many times as there are genes, to arrive at 2^{40} possible phenotypes. This number is already greater than a trillion, but it grows much larger if we consider that a gene can be more than just on or off—it can express a small, medium, large, or very large amount of regulator protein. What's more, several circuits often cooperate to shape any one body part, which multiplies the number of possible expression codes.[46] Compared to the number of these possible meanings, the few hundreds of cell types and tissues of a complex body like ours are paltry. Even if we allow that all cells in a body must be laid out with precision in space, there are plenty of expression codes to go around—perhaps too many to find any one of them.

Evolution explores this circuit library through the familiar crowd of randomly browsing readers, populations of organisms in which circuits get modified through occasional DNA copying errors that arise as genes get passed from parents to children with some corrupted letters. Any one such mutation can have two kinds of effects. It can deform a regulator's shape and prevent it from recognizing DNA. Or it can alter one of the DNA "words" a regulator recognizes, either clipping a single wire of the circuit—actually disrupting the regulator's effect on a gene—or creating a new wire, a new molecular word recognizable by some regulator.

The first kind of change often results in disaster, because each regulator affects so many other genes. Destroying a regulator's ability to recognize DNA is akin to scrambling a complex recipe's ingredients and thus destroying the entire dish. It can lead to organisms with terrible malformations or to embryos that die before they are born. The second sort of copying error, however, is more like a typo in a recipe. By changing the activity of merely one gene and the amount of protein it expresses—one among thousands of protein ingredients—it is less likely to cause serious damage. One might think that changes of this second kind are more tolerable and could thus steadily accumulate on evolutionary time scales. If so, they could slowly transform a circuit's wiring diagram.

When one compares circuits that have evolved separately over millions of years, like those in some of the more than a thousand different fruit fly species, one finds indeed that most tolerable changes have occurred in the wires and not in the circuit genes themselves. Evolution alters most circuits one wire at a time, because messing with the circuit genes invites disaster.

What is more, these small wiring changes indeed accumulate to transform circuits, and this process is far from slow.[47] The reason is that a regulator's DNA keyword can be as short as five letters and occur thousands of letters away from a gene. By chance alone, random mutations can easily create new keywords and thus new wires in a circuit.[48]

If only one and no other circuit in the hyperastronomical library of 10^{700} texts expressed the code to create a specific innovation, evolution might as well pack up and go home, because this code would be a needle in a haystack many times the size of the universe.[49] The question of why it had not thrown in the towel had called to me since the early 1990s, but I had ignored it—too many other projects. My procrastination ended in 2004 when I spent a research sabbatical at the Institute for Higher Studies near Paris in France.[50]

Set in a bucolic park with plenty of old trees, sculpted shrubbery, overflowing flowerbeds, and footpaths to explore while pondering life's questions, the institute is a monastic refuge from the endless fund-raising, networking, and community service of an academic's life. Its few resident researchers are highly decorated scientists, among them several recipients of the Fields Medal, widely known as the Nobel Prize for mathematicians. The institute focuses on mathematics and physics, but its leaders were aware that a seed long dormant in molecular biology, the insight that the whole is more than the sum of its parts, had germinated and flowered into an enormous subdiscipline known as *systems biology*. This emerging research field joins experimental data with mathematics and computation to find out how molecular parts like a fly's regulators cooperate to shape whole biological systems, that is, organisms.[51] Mathematicians and physicists have many tools to crack problems like this, and so the institute invited biologists like me for extended visits to see what we could do together.

Luckily for me, I accepted the invitation. Because it was in Paris that I met Olivier Martin.

Olivier is an internationally respected professor of statistical physics at the University of Orsay near Paris. Statistical physicists like him deal with huge collections of things like the molecules in a pressurized container of propane gas, and how they create properties like the pressure of that gas. To predict this pressure is important—we don't want our gas tanks to explode—but also impossibly complex, because trillions of molecules bounce into the container walls every instant. Statistical physicists love to think about wholes with trillions of parts—too many to track individually—and they develop clever ways to describe those wholes, employing sophisticated statistical methods that share little beyond the name with the statistics that pollsters use to predict the outcome of elections in the United States.[52]

Olivier had a problem, though. Statistical physics is like a buffet where a hungry mob has devoured the choice dishes and left only small morsels behind: Most of its big questions are answered, and the remaining ones are either too hard or too trifling—not altogether surprising since scientists like James Clerk Maxwell and Ludwig Boltzmann had solved thermodynamical problems with statistical methods since the nineteenth century. Like most scientists in his situation, Olivier wanted to make a bigger contribution than physics would allow him. His problem was to find a question in systems biology that was new and challenging enough for him to chip away at.

I had a library of 10^{700} regulation circuits to map. Boy, could I help Olivier out.

As Olivier Martin and I began to collaborate, I first came to appreciate him as a scientist whose intuition and technical skills prevented us from getting lost in the library. But he turned out to be much more than a sure-footed travel companion. He was a kind and generous teacher who would patiently explain how the tools of his trade could help us find our way.[53]

We started with small steps whose purpose was to answer a question you will recognize. Is there only one text in the circuit library that expresses any one meaning? To find out, we started out with a single circuit in the library and computed its expression code. Then we changed one wire, asked whether this mutation altered this expression phenotype, went back to our starting circuit, changed another wire, and so on, until we had created all neighbors of our circuit and knew their phenotypes. And to make sure that the neighborhood of this one circuit was not unusual, we explored the neighborhoods of many different starting circuits, circuits with different numbers of genes, different numbers of wires, a different arrangement of wires, and different phenotypes.

They all gave the same answer. Circuits typically have dozens to hundreds of neighbors with the same phenotype.[54] In other words, the phenotypes of these circuits remain unchanged even after encountering mutations that alter individual wires. They are not quite as delicate as those acrobat-formed human sculptures, where a body's shifting by a few millimeters can spell disaster. Regulatory genotype circuits can tolerate such changes, because not all individual wires are critical to their function.

This first step away from a circuit already told us something very important: No one expression code—be it the one segmenting a fruit fly, dissecting a leaf, or shaping a vertebral column—has only one, special, unique circuit producing it. Each expression code can be produced by many circuits that differ in how their genes are wired. Finding out how many was trickier, because the number is so large that we could not even compute it, at least for circuits of forty or more genes. All we knew was that the number had to be enormous, since we had been able to calculate it for smaller circuits: Those with ten genes already had more than 10^{40} circuits, and those with twenty genes had more than 10^{160} circuits able to produce a given expression code. Producing any one expression code is another problem with more solutions than one can count.[55]

To find out how far apart different solutions to the same gene expression problems are in the library, we took the same random walks we had used when investigating metabolisms and proteins. Starting with a circuit, we computed its expression code, altered a wire—adding or eliminating regulation of one gene—and thus stepped to a random neighbor with the same expression code, and from that to the neighbor's neighbor, and so on, until we could not go further without changing the expression code.

Once again, we could walk almost all the way through the library. Circuits that differed in more than 90 percent of their wires could still produce the same expression code. Looking at their wiring diagram, you would never guess that one arose from the other in many tiny steps. Yet each one was a different solution to the same problem: how to produce a specific pattern of gene expression that can shape a cell's identity.

To make sure that the starting circuit—and its expression code—was not unusual, we started to explore the library from many different shelves, circuits with different numbers of genes, different numbers of wires, different arrangement of wires, and different expression patterns. It did not make a big difference. Some circuits with the same expression code differed in every single wire, whereas others differed in "only" 75 percent of their wiring. But even these would not be recognizably related when examined side by side.

Our explorations also taught us that *all* circuits with the same expression code are typically connected in the library. We can start from any one of them, change one wire at a time, and transform the circuit step by step into *any* other circuit with the same meaning, such that each step leaves the meaning unchanged.[56] Once again, we could find a path from nearly every point in the library to nearly every other one, without ever getting stuck in a morass of regulatory nonsense.

All this means that circuits with the same phenotype form a vast network in the library of circuits, a *genotype network* like those we found in the metabolic library and the protein library. The library is filled with these networks, each of them containing more circuits than you can count, each of them reaching far through the library. All circuits in the same network are solutions to the same problem: how to produce a specific expression code that helps shape a cell, a tissue, or an organ. Small wonder that innovations like dissected leaves could evolve dozens of times independently, if vast numbers of circuits have the expression code that can get them there.

To map the millions of circuits needed to understand the library would have been impossible with any available technology aside from computation—hundreds of researchers had to experiment on millions of fruit flies over several decades to understand the single circuit segmenting a fly. However, some intrepid scientists are beginning to map circuits in simpler organisms, such as bacteria and yeasts. One of them is Mark Isalan, a researcher in Barcelona, who rewired a transcriptional regulation circuit in *E. coli* by adding new wires—regulation between pairs of genes—and created hundreds of circuits in its neighborhood. And he found, as we had, that regulation circuits are sturdy enough to be rewired.[57] Ninety-five percent of his rewired circuits function normally.

Other researchers compare regulator circuits among different species of the yeasts we use to brew beer, to see how far they have to travel through the circuit library. One such circuit activates genes that allow yeasts to digest the sugar galactose. You might think that there must be one *best* way to wire this circuit, and that the yeast species that has discovered this way would have passed it on, unchanged, to others. Not so. In two yeast species that split many million years ago, this circuit not only has become completely rewired but even uses different regulators.[58] Neither of these circuits is inferior, otherwise it would not have survived. Nature has solved the same regulation problem in two different but equally adequate ways. Not only that, but a path of small mutational steps connects these solutions, because the species shared a common ancestor.

Genes for the ribosome, the complex multiprotein machine that translates all RNA into proteins, tell the same story. A cell must manufacture its dozens of proteins in precisely balanced amounts, or else it will disappear like those wasteful *E. coli* cells overproducing beta-gal. Achieving this balance might seem a delicate affair with only one best solution, but again, two different species of yeast have come up with equally successful solutions that regulate these genes in completely different ways.[59]

Examples like these show that organisms can indeed travel far through the circuit library. But when searching for rare nuggets of new and useful expression codes on their journey, they face a problem similar to that of innovating metabolisms and proteins: There are many trillion possible expression codes, but the immediate neighborhood of any one circuit contains at most thousands of other circuits—those differing in one wire—too few to find all possible expression codes nearby. To discover myriad new expression codes, evolving circuits need to venture out of their neighborhood. Such expeditions yield many discoveries only if *different* neighborhoods contain *different* expression codes. To find out whether they do, we asked our computers to draw two arbitrary circuits from the same genotype network—call them A and B, they produce the same expression code but have different wiring—identify all circuits near them, and compile a list of expression codes of all these circuits. We found that most expression phenotypes in the neighborhood of A are different from expression phenotypes in the neighborhood of B— regardless of A and B's phenotype, number of genes, or wiring. Different neighborhoods contain different phenotypes.

So we are back to a familiar story. The regulatory circuit library has the same layout as the metabolic and the protein libraries. Circuits with the same gene expression phenotype are organized in vast and far-reaching genotype networks. And that has consequences for a crowd of readers wandering aimlessly along such a network, figuratively seeking something new to read, actually propelled only by the steady if direction-less force of mutation that slowly changes circuits, one regulatory inter-action at a time: Even though some steps garble a circuit's expression code, many others preserve it and thus allow readers to move along the genotype network. While the readers wander, they reach ever-new neigh-borhoods that contain texts with ever-new meanings, ever-new expres-sion phenotypes, one of which may seed the next big thing in life's architectural contests. Once again, genotype networks and their diverse neighborhoods create innovability.[60]

These similarities among different libraries are mysterious. How could innovability in metabolism, in proteins, and in regulation circuits have the same source, a library full of chemical meaning with a common cataloging system? The answer is held by an invisible hand that guided the world long before life's origin—self-organization, a peculiar kind of it. We will turn to it next.

CHAPTER SIX

The Hidden Architecture

I n 1944, the Nobel Prize–winning theoretical physicist Erwin Schrödinger published a series of his lectures under the title *What Is Life?* The brief book was an attempt to reconcile physics with what was then known, in the days before Watson and Crick, about evolution. The book is brimming with ideas, and one of them has spilled into the mainstream of popular science culture: It is the idea that evolution increases order and decreases disorder—what Schrödinger called "negative entropy." Four years later, the American electrical engineer Claude Shannon connected the thermodynamic concept of entropy to the problem of transferring information through a telegraph line. The concepts of evolution and information have been linked ever since, usually in a fairly primitive way. Disorder: bad. Order: good. Positive entropy: bad. Negative entropy—now also known as information—good.

In the years since Schrödinger's book, we have become more sophisticated in thinking about entropy. Order and information remain central to evolution, but in recent years we have also learned, thanks to genotype networks, that perfect order is as hostile to innovation as total disorder. Nature doesn't just *tolerate* disorder. It *needs* some disorder to discover new metabolisms, regulatory circuits, and macromolecules—in short, to innovate.

Let's put Lego blocks to another metaphorical use, and consider the difference between a disorderly jumble of those familiar plastic tiles and an arrangement in which every tile has been presorted into a "proper" place, and where a child must assemble them in a specific sequence to build a pirate ship following a plan helpfully provided by Lego. The disordered jumble of Lego tiles has greater potential for innovation than the carefully organized one, and not just because it stimulates a child's natural creativity to find new ways for building pirate ships. A deeper reason is that there are many more ways to build a pirate ship than those contained in Lego's instruction book.

In biology this simple fact is manifest in the multiple solutions that nature found—courtesy of genotype networks—for problems like protecting organisms against freezing. And it is also deeply connected to a biological phenomenon little appreciated until the end of the twentieth century, but in fact so widespread that it deserves to be called a hallmark of life: *robustness,* the persistence of life's features in the face of change.

The meaning of robustness is best illustrated with the difference between typographical mistakes in a traditional book and in a computer program. A book containing the letter sequence

N smll stp fr mn, n gnt lp fr mnknd

would raise eyebrows, but the meaning of this sentence remains understandable. However, a single misplaced letter or as little as a missing comma in a thousand pages of computer code can bring a million-dollar software package to a crash. Software bugs like this cause billions of dollars in economic loss every year. Human language is robust. Programming language, not so much.

The suspicion that life is robust arose at least as far back as the 1940s, when the biologist and philosopher C. H. Waddington studied flies with different genotypes and discovered that they had indistinguishable bodies, down to the minutest details of their wings' venation and the numbers of bristles that cover their backsides. He called the phenomenon through which development can produce "one definite end-result regardless of minor variations in conditions" *canalization*—another word for robustness.[1] And although his research hinted that the body plans of flies are robust to genetic change—there are many ways to build a fly's body—research into robustness remained a backwater for another half century.

But almost overnight in the 1990s, robustness entered center stage with a flourish when molecular biologists were baffled by a discovery superficially unrelated to Waddington's: Many genes apparently serve no purpose.[2]

What's baffling about such genes is why they would exist at all. Not only does a superfluous gene waste scarce resources, but mutations that incessantly rain down on DNA would eventually erode it, transforming it over time into something like an abandoned building that crumbles to dust over the years.[3]

Many of these "purposeless genes" were unearthed after the genome of an organism that we already met in chapter 5 had become fully sequenced. It is a microbe, the brewer's yeast *Saccharomyces cerevisiae* that helps us make beer and wine, and it is as useful for understanding cell biology as fruit flies are for embryology.[4] The yeast genome in hand, biologists realized that thousands of its genes had an unknown role in the microbe's life. To reveal this role, they began to engineer "knockout mutations" into the genome, so called because they delete a single gene, an entire meaningful paragraph from a genomic text.[5] The logic of the experiment is essentially like analyzing the workings of a car by eliminating one part at a time: Remove the disk rotor, and if stepping on the brake pedal no longer slows the car, you have learned that the rotor is needed for braking. In the same way, if you knock out a particular yeast gene and find that its cells can no longer divide, the gene was involved in cell division. Knock out a fruit fly gene and if the mutant no longer forms wings, you know that the gene helps build wings.

The results of gene knockout experiments had trickled into the scientific literature gene by gene, until gene-knockout technology became powerful enough to delete thousands of genes. That's what researchers at Stanford did in impressive experiments starting in the late 1990s, when they used the list of yeast genes revealed by the yeast genome and set out to delete every single yeast gene. They created some six thousand different yeast mutants, each of them missing a gene, placed these mutants into chemical environments where their unmutated ancestor could have thrived, and examined each mutant for specific defects, clues about the missing gene's function.[6]

What they found was completely unexpected. Thousands of these mutants do just as well as their ancestor and show no obvious defects. The genes that had been deleted to create these mutant genomes served no obvious purpose. Since then, scientists have blocked countless genes in many other organisms. And these genes tell the same story as a vowel-free English sentence: Like natural language, life is robust—in this case to gene deletions.[7]

Discoveries like this do mostly one thing: They create new questions. One of them was *how?* What mechanism creates robustness?

For some genes the mechanism was straightforward. These genes were duplicates, stretches of DNA that occur more than once in a genome, like pages in a book that someone has photocopied twice by mistake. Gene duplications happen when an organism copies or repairs DNA, and are by no means rare: About half of the genes in our own genome have duplicates.[8] Since identical duplicates can do the same job, one of them can take over if you knock out the other.[9] Like the redundant power supplies that hospitals use to safeguard against power failure, like redundant computer memory to prevent data loss, like redundant circuitry in commercial aircraft to prevent crashes, some genes are only "useless" until they're needed.[10]

But many of the dispensable genes have no duplicate—they are single-copy genes—and for them the causes of robustness are not as simple.

We understand those causes best for genes that encode the enzymes of metabolism. A metabolism's chemical reaction network resembles the dense road network of a city's core, like that formed by the right-angled streets of midtown Manhattan. A driver who wants to get from 42nd Street and Second Avenue to 48th Street and Seventh Avenue has any number of choices for following the street grid six blocks north and seven blocks west. The major arteries have multiple lanes—think of them as redundant, because even if one is blocked, the driver can continue in another. But even a complete roadblock is not a problem, because the driver can use a different part of the grid, and a really intrepid driver might even cut through parking lots with entrances on two parallel streets. Such detours slow down but don't halt the journey.

A knocked-out metabolic gene is a bit like a blocked road that halts the flow of molecules through a network of metabolic reactions. The detours around the roadblock are alternate metabolic pathways, sequences of chemical reactions that can absorb the backed-up molecular traffic, synthesize needed molecules in different ways, and ensure that life in metabolism city goes on.[11] This isn't just an abstract metaphor. Biotechnologists can create metabolic roadblocks by knocking out metabolic genes, and when they do, organisms like brewer's yeast often survive by rerouting the flow of essential molecules. In metabolism, this kind of robustness is even more important than redundancy.[12]

Robustness isn't limited to metabolism and whole genomes. It is just as pervasive in individual proteins like *lysozyme*. This protein kills bacteria by destroying their wall of protective molecules. It appears not only in human saliva, tears, and even mother's milk, but in a large number of other animals, and even in viruses that attack bacteria.[13] When scientists want to find out how a protein like this works, they do something akin to knocking out genes in a genome, but on a smaller scale—they change individual letters in the protein's amino acid string and observe the effects of each change. When they engineered more than two thousand lysozyme variants, each one with a single altered amino acid, they found that some sixteen hundred variants—more than 80 percent—could still kill bacteria. Proteins like lysozyme, and there are many, are as robust as metabolisms. And the same holds for regulation circuits—we already heard about a circuit in the bacterium *Escherichia coli* that can be rewired in the laboratory without ill effects (chapter 5).

The most obvious benefit of such robustness is that it keeps organisms alive. Its importance goes back all the way to the first self-replicating RNA molecules and the fatal error catastrophe, in which small errors compound over time until replication becomes impossible (chapter 2). That was a true catch-22: RNA molecules have to self-replicate with few errors to acquire the ability to self-replicate with few errors. But only a bit of the robustness in today's RNA could lower the bar for this problem to a manageable height: Because a few replication errors in a robust molecule do not erode its ability to self-replicate, robustness provides a stay of execution by error catastrophe, perhaps long enough to stumble upon better replicators.[14]

But the importance of robustness goes far beyond that. It explains the mystery of genotype networks and of innovability.

To see why, we need to revisit nature's libraries, where each metabolism (or protein, or regulation circuit) is represented by a single text, and where each of this text's neighbors differs in a single letter, a single reaction, or a single enzyme and its gene. We know from gene deletion experiments that many of these neighbors, for example metabolisms where a single reaction has been eliminated through a gene knockout, suffer no ill effects. This means that even when the *genotype* has changed, there need not be any change in the *phenotype,* in the organism itself and its observable features. An organism like this is robust. The extent of its robustness is reflected in the number of its neighbors—variants a single small change away—whose phenotype remains unaffected by the change: The more neighbors with the same phenotype, the more robust the organism is.[15] Think of this phenomenon at its theoretical limits: If a metabolism, or a protein, or a regulatory circuit had no viable neighbors it would be maximally fragile. Change one of its parts, and death follows. At the other extreme, if every possible change were viable, if every neighboring metabolism had the same phenotype, the metabolism would be maximally robust: No single change could kill it.[16]

These extremes do not exist in the real world. No real organism completely lacks robustness, and no organism is perfectly robust. But all organisms, their structures and activities, are *to some extent* robust, and it is precisely this robustness that allows populations to explore nature's enormous libraries. The number of texts with any one meaning in these libraries is vast, but these texts fill a tiny fraction of the library, like a droplet of molecules in an ocean. In the complete absence of robustness, many texts might tell the same story, but none of their neighbors would. No explorer could browse one text and find a neighboring one with a single page—or word, or letter—changed but its meaning nonetheless intact. Genotypes with the same phenotype would be like stars in the sky—a billion twinkles isolated by light-years of empty space.

Luckily, the biological world is different. Starting at any one robust text, we can step to one of its many neighbors with the same meaning, and we can step to one of *its* robust neighbors, and so on, never changing this meaning, and thus exploring ever-new regions of nature's libraries that harbor untold innovations. Robustness allows *some* disorder in genotypes, and permits nature to explore new configurations of its Lego blocks through the genotype networks it helps create.

Genotype networks are yet another example of the pervasive

self-organization we first encountered in chapter 2—the same phenomenon that pervades both the living and nonliving worlds, from the formation of galaxies to the assembly of membranes. But they are a peculiar
example of self-organization. Unlike galaxies, which self-assemble
through the gravitational attraction of cosmic matter, or biological membranes, which self-organize through the love-hate relationship of lipid
molecules with water, genotype networks do not emerge over time. They
exist in the timeless eternal realm of nature's libraries. But they certainly
have a form of organization—so complex that we are just beginning to
understand it—and this organization arises all by itself. And as with galaxies and membranes, the principle behind their self-organization is
simple: Life is robust. This robustness is both necessary for genotype
networks—otherwise synonymous texts would be isolated from one
another—and it is sufficient.[17] Wherever metabolisms, proteins, and regulatory circuits are robust, genotype networks emerge.

Robustness is sufficient to create genotype networks, but genotype networks alone are not sufficient for evolution. The reason is that evolution
must meet two demands, seemingly at odds with one another. It must be
simultaneously conservative and progressive, like some aviation pioneer
embarking on a transatlantic flight in the Wright brothers' original flyer:
Certain in the knowledge that he must invent a new design to complete
the journey, he must also keep the old one in the air until he does. Nature
must keep what works alive while exploring the new. Genotype networks
are essential for exploration. But they aren't made for conservation.

This bears emphasizing, because the exciting new discoveries about
genotype networks can tempt us to forget the critical importance of natural selection. Conservation is the job of natural selection—evolution's
memory—and its power to conserve even tiny improvements, given
enough time, is so great as to seem absurdly unbelievable. Literally so. In

the *Origin* Charles Darwin wrote about eyes, surely among the most spectacular innovations in life's history, "To suppose that the eye with all its inimitable contrivances for adjusting the focus to different distances, for admitting different amounts of light, and for the correction of spherical and chromatic aberration, could have been formed by natural selection, seems, I freely confess, absurd in the highest degree."[18]

When light passes through our eye, the lens projects a fantastically accurate, undistorted image of the outside world onto our light-sensing retina. To do so, it must *refract* light's path, changing its direction at a precise angle.[19] It's not just the lens's shifting shape that makes this possible, but also the less appreciated and peculiar lens material—an ancient innovation that required nothing but new regulation.

Shine a flashlight obliquely onto a body of water, and you will see that the light ray kinks at the surface. Dissolve sugar in the water, and the kink's angle gets sharper—the more sugar you dissolve, the sharper it gets. (The food industry uses this principle to measure the amount of sugar in wine, soft drinks, and juices.) Our eyes refract light just like that, except that they use proteins instead of sugar. These proteins—crystallins—occur at very high concentrations in the lens, which allows the lens to refract light strongly.

Crystallins are so uncannily good at refracting light that it's tempting to think that they were tailor-made for constructing lenses, and therefore rare. Not true. Many crystallins are metabolic enzymes, the very same enzymes that promote chemical reactions elsewhere in the body, albeit in smaller numbers. Different organisms use different enzymes as crystallins. What distinguishes them from other proteins is that they do not clump easily, not even when they are expressed at the extreme concentrations needed in the eye.[20] Eyes build their lenses out of proteins like the one that detoxifies alcohol, just because they confer transparency—the same way you might use an old brick as a bookend simply because it happens to be heavy. Crystallins are also some of the sturdiest proteins around, so long-lived that the crystallins that make up the lens of the human eye last an entire lifetime, from birth to death.[21] But sometimes they wear out and start to clump, making the lens milky white. When this happens a cataract has formed, with consequences both well known and disastrous: blindness.[22]

Though Darwin knew nothing of protein chemistry, he did suspect—and today we know—that the fancy eyes of vertebrates with their sophisticated lenses are the last in a long list of gradual improvements. Long before our ancestors started co-opting nonclumping metabolic proteins, *their* ancestors, like some worms and starfish, were using flat patches of light-sensitive cells that were at least good enough to find a shadow to cower in and hide from predators. After millions of years, these cells eventually congregated in shallow bowls, *eyecups* that can detect light's direction better, which deepened into *pit eyes* that can detect it very well, and even further into *pinhole* cameras whose tiny openings can produce real images. From there it was one more step to lenses, transparent tissues of higher density that—thanks to crystallins—could focus light. Eventually these lenses became able to flex or move to create sharp images.

All these small, gradual improvements are worth preserving, and natural selection did. We know, because many animals still have them: eyecups in some flatworms, pit eyes in some snails, pinhole camera eyes in the nautilus—a relative of squids that builds many-chambered shells—and simple lenses in organisms as primitive as jellyfish.[23]

It's a bit like the stunning grandeur of medieval cathedrals, with soaring spires, columns of heavy massive stones assembled with exquisite precision, and vaulted ceilings so high that our gaze gets lost in their semidarkness. The finished product—like the human eye—is literally incredible without the knowledge that it was built one brick at a time.

The same is true for all molecular innovations. The amino acid text of the Arctic cod's antifreeze proteins didn't originate in a single step, like Athena springing from the brow of Zeus. But every single letter of an ancestor's amino acid text that changed in the right direction, lowering a body fluid's freezing point by as little as a tenth of a degree, could expand its descendants' habitat by miles. A greater range means a larger and more varied food supply. It means a change well worth preserving, and a long sequence of such tiny changes can expand life's frigid frontier by long distances. Genotype networks are crucial to find each such change, and natural selection is crucial to preserve it.

Better variants that improve an organism incrementally are important for innovation, but they are not the only kind of change that DNA experiences. Many mutations neither harm nor help when they first arise. Such *neutral* changes are a consequence of life's robustness and the disorder it allows.

That neutral changes could matter for innovation—and why—was not always clear. In fact, the relationship between natural selection and neutral change was central to a historical controversy that tore at the fabric of Darwinism in the last third of the twentieth century. The revolution in molecular biology, then well under way, had revealed that populations of wild organisms, from mammals to fruit flies and down to bacteria, harbored astonishing amounts of genetic variation: The DNA of thousands of genes in members of the same species varied in its letter sequence. Most scientists, good Darwinians as they were, believed that the fate of most of these variants was determined by natural selection—variants that appeared frequently must improve survival or reproduction.

But these selectionists were opposed by a vocal minority, the neutralists, who argued that most of these variants make no difference to the organism and are invisible to selection. At least when they first appear, they are neutral. In the eyes of some, like the paleontologist Stephen Jay

Gould, the very existence of neutral change compromised the importance of natural selection in evolutionary innovation.[24]

The history of science and technology offers loose analogies for how neutral changes—dormant discoveries—can become valuable for future innovations. Number theory provides one such analogy. It is a branch of mathematics about which the American mathematician Leonard Dickson reportedly said, "Thank God that number theory is unsullied by any application."[25] This was as true in 1919 as it had been since Euclid, but within decades unrelated developments—digital computers and networked communication between them—placed the theorems of number theory at center stage of the Internet economy, where they ensure the secure communications that make e-commerce and online banking possible. In a similar vein, the German physicist Heinrich Hertz, whose experiments validated the electromagnetic theory of James Clerk Maxwell, saw no practical purpose to his discovery. He reportedly said that it was "of no use whatever" and "just an experiment that proves Maestro Maxwell was right." Less than forty years later, his discoveries led to the first commercially licensed radio station in the world—KDKA in Pittsburgh, which still broadcasts on the frequency band of 1020 kilohertz.[26]

But back to biology, where the neutralists' most outspoken proponent was the Japanese scientist Motoo Kimura, who had developed a sophisticated and successful mathematical theory to explain the evolutionary fate of such neutral mutations. Kimura asserted that *most* genetic variation seen in nature is neutral. The genomic era has taught us that he was wrong on that count—neutral variants are no more frequent than those providing an advantage. However, his hunch of neutral change's being important was dead on, though it took another few decades to understand why.

One reason is that neutral change is critical for navigating genotype networks. Neutral change provides the browsers of nature's libraries with a safe path to innovations through treacherous territory of meaningless texts. Without genotype networks and the neutral changes they allow, the exploration of nature's libraries would be just about impossible.

Another reason is that a change that is neutral when it first appears doesn't have to stay that way. Once-neutral changes can turn into essential parts of innovations—like the theorems of number theory. And once they do, natural selection can preserve them. Which means that both selectionists and neutralists had a point, because both kinds of changes are essential in evolution. After neutral changes have paved the way to an innovation, selection preserves those neutral changes that contributed to the innovation.

A single example from a well-studied RNA enzyme known as the hammerhead ribozyme can illustrate to what extent neutrality and genotype networks can accelerate evolution's search for innovations. The job of this particular RNA enzyme is to cleave RNA at a particular point along its nucleotide strand. The shape of the ribozyme, named for its apparent resemblance to the hammerhead shark, is what allows it to perform this job—adequately but not necessarily optimally. Somewhere out in the vast library of all RNA molecules might be new shapes, new phenotypes that would endow the ribozyme with sharper blades.

If no genotype networks existed, then a crowd of readers in the RNA library—an evolving RNA population—would have to congregate around the forty-three-letter-long text that encodes the RNA, and could only explore the shapes that are one letter change away. This particular RNA enzyme happens to have 129 neighbors, and because we can compute their shapes, we can determine that there are forty-six new shapes in this neighborhood.[27] That's the number of shapes evolution can explore without genotype networks.

And with them? If we only step to the text's neutral neighbors—those with the same hammerhead shape—and determine the shape of all *their* neighbors, we already find 962 new shapes. And if we just walk one step further, to *those* neighbors' neutral neighbors, we find 1,752 new shapes. Just two steps along this ribozyme's genotype network, we can access almost forty times more shapes than in its immediate vicinity. The genotype network of the hammerhead shape of course extends much further than just two steps, and it has more than 10^{19} members, too many to count all the new shapes near them with current computers.[28] But we can say with certainty that countless millions and billions of new shapes are near them, all explorable because evolution's readers can spread out along the genotype network without suffering death.[29]

That's how much genotype networks accelerate innovation. They are like the warp drives of *Star Trek,* science fiction's solution for faster-than-light interstellar travel. Extrapolate from the hammerhead ribozyme and imagine that evolution had unfolded merely forty times more slowly without genotype networks. Instead of being four billion years along its path, life would have evolved only as far as it did during the planet's first hundred million years. A few kinds of bacteria would be around, but certainly no multicellular organisms, let alone fish, land plants, dinosaurs, or nonfiction book authors. And genotype networks accelerate evolution much more than fortyfold—so much more we can't even yet compute it. Without them, life would never have crawled out of the primordial soup.

Science fiction also has another solution to faster-than-light travel: changing the shape of space itself. Inventive science-fiction authors have postulated technologies like "wormhole drives" that allow instantaneous travels between places thousands of light-years apart. It turns out that genotype networks also do something similar. They shrink the distance between texts in the libraries of metabolisms, macromolecules, and regulatory circuits.

Imagine a crowd of evolution's readers—organisms in a population—congregating near a text describing a circuit with a specific expression code that helps shape some body part, like a bird's wing. Now imagine that somewhere in the library of regulatory circuits a *new* code exists that modifies the wing to make it slightly more aerodynamic or lighter. The further the readers must travel to find it, the more time they need to find this innovation.

Browsing through such an enormous library seems, at first glance, like hunting for a particular needle in a haystack. You might find the right needle immediately, but chances are you would have to examine most of the haystack—perhaps all of it—before achieving success. Common sense dictates that the same applies to the library, that a new expression code is like a single needle in a haystack many times the size of the universe.

But common sense fails in this library. We had already learned this from our discovery that there are innumerable circuits with the same expression code—the haystack has many needles—but the library is even more bizarre, as we found out by searching for circuits with specific new expression codes. In this search, we created arbitrary expression codes—thousands of them—and for each such code we used our computers to generate a pair of circuits, where the first circuit produced one of these codes, and the second produced the other. The two circuits differed also in most of their wiring pattern, in the who-regulates-whom among the circuit genes. We then changed the first circuit gradually, one wire at a time, requiring that each such gene regulation change preserve the circuit's expression code. How close could we get to the second circuit? Very close, we found out, for example to within 85 percent of the second circuit's wiring for circuits of twenty genes. In other words, starting from anywhere in the library—*anywhere*—you need not walk very far, only fifteen steps away from a genotype network, before finding the genotype network of *any* other circuit. It is as if your needle were *always* nearby, no matter where you started to search.[30]

If this doesn't sound strange enough, get ready for something even stranger.

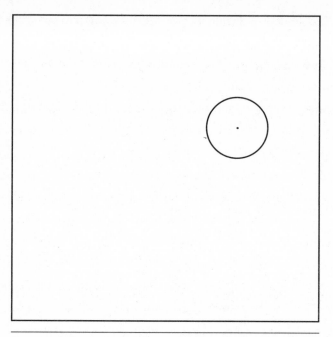

FIGURE 19. The square stands for the entire library, the circle for a neighborhood of some genotypic text (dot) in the library.

Imagine that the square in figure 19 is the library, and the dot is a single text in it. The circle around this dot has a radius that is 15 percent as long as the sides—this is how far a reader would have to travel from a genotype network, on average, before finding a specific new expression code, as we had found out from exploring the circuit library. A simple calculation shows that a circle with a radius of fifteen centimeters, inside a square that is one hundred centimeters on a side, has an area of 707 square centimeters, a little more than 7 percent of the square's area.

Actual libraries aren't two-dimensional, of course. They exist in three dimensions. For simplicity's sake, let's say that our library building is housed in a three-dimensional cube, and the area of the library containing circuits with a new expression code is now a sphere. If so, then the sphere inscribed inside the library will now have a radius 15 percent as long as the sides of the cube, just as with the square. The ratio of their volumes, however, becomes very different. The volume of the sphere covers not 7 percent of that of the cube, but only 1.4 percent.

The libraries of regulatory circuits, of course, aren't three- or even four-dimensional. They occupy higher dimensions, where cubes become hypercubes and balls become hyperspheres. In four dimensions, our hypersphere—still with the same radius—contains 0.2 percent of a hypercube's volume. In five dimensions, it covers 0.04 percent, and so on.

In the number of dimensions where our circuit library exists—get ready for this—the sphere contains neither 0.1 percent, 0.01 percent, nor 0.001 percent. It contains less than 0.00000000000000000000000000000 0000000000000000000000000000000000000000000000000000000000 0000000000000001 percent, or one 10^{-100}th, of the library.[31] This is the tiny fraction of the circuit library a reader needs to explore, starting from its own genotype network, to find a circuit with any one expression code. That it is so small emerges from a simple geometric principle in spaces of ever-higher dimensions: A ball with constant radius occupies ever-decreasing and ever-tinier fractions of a cube's volume.[32] (This volume decreases not just for my example of a 15 percent ratio of volumes, but for any ratio, even one as high as 75 percent, where the volume drops to 49 percent in three dimensions, to 28 percent in four, to 14.7 percent in five, and so on, to ever-smaller fractions.)[33]

The same counterintuitive phenomenon holds in other libraries of innovation: The more dimensions they have—the larger their collection of metabolisms or molecules—the smaller the distance to find specific innovations. A browser who starts with a metabolism that can survive on some foods and then blindly searches for one that can thrive on others needs to change only a few reactions and explore a tiny fraction—too small to imagine—of the metabolic library before stumbling upon the right text. The same holds true for RNA. Starting from an existing RNA molecule, a nearby molecule with a new shape—*any* new shape you choose—will be found after changing only a few of the molecule's nucleotide building blocks and having explored a tiny fraction of the library.[34]

The astonishing fact that evolution needs to explore one 10^{-100}th of a library to secure the arrival of the fittest goes a long way to explain how blind search produces life's immense diversity. Evolution does not have to search the entire haystack, because the haystack contains more than one needle. In fact, thanks to robustness, and the genotypic disorder it permits, the haystack contains too many needles to count, and they are organized into sprawling but navigable networks.

And if we remember that the neighborhood of each text is extremely diverse, we have understood one more feature of the library's organization: Genotype networks not only range far and wide but also are tightly interwoven. They form a dense tissue of networks, each genotype network surrounded by many others, interwoven with them on all sides, a tissue so complex that it looks nowhere the same, consisting of millions, billions, or more of different strands, each one corresponding to a different phenotype. If each strand had a different color, this tissue would be woven in such an intricate way that near any one strand, threads of billions of other colors would pass. Only a high-dimensional space can host such a fabric, whose texture is intricate beyond our grasp. This fabric is different from anything we know. It is hidden behind the visible splendor of each living thing, yet all this splendor emerges from it.

Because genotype networks and their fabric are a consequence of robustness, robustness is immensely valuable to innovation. But valuable things are usually not free, and robustness is no exception. Its price—a high one—is complexity.

It's almost too easy to criticize complexity. In Lewis Carroll's *Through the Looking-Glass*, Alice is navigating the tale's fantastic chessboard when she is attacked by the Red Knight, who mistakes her for a white pawn. In the nick of time, Alice is rescued by his opposite number, the White Knight—tellingly, an inventor—who is eager to show his new friend his latest innovations, including boxes that open on the bottom to keep out the rain, a device for trapping mice when they appear on a horse's back, and a dessert made of blotting paper, sealing wax, and gunpowder:

> "You see," he went on after a pause, "it's as well to be provided for *everything*. That's the reason the horse has all those anklets round his feet."
>
> "But what are they for?" Alice asked in a tone of great curiosity.
>
> "To guard against the bites of sharks," the Knight replied. "It's an invention of my own."

The White Knight is so handicapped by his complex and fantastic inventions that he is literally unable to ride and to accompany Alice on her journey, and so he quickly departs the story. He lives on, however, as an object lesson in the importance of simplicity.

Long before Carroll wrote his story, the fourteenth-century English friar William of Ockham had already expressed an enthusiasm for simplicity when he coined a now famous principle of parsimony: that phenomena should always be understood using the fewest possible facts, or "entities," as Ockham called them. This ideal—often called Ockham's (or Occam's) razor—is usually held up to scientific explanations that are supposed to be both truthful and beautiful. But it could apply equally to the kinds of machines inventors and engineers build, even though engineers already have their own, more earthy motto: KISS, for Keep It Simple, Stupid.

The ideal of simplicity is not just an aesthetic ideal or a philosophical principle. In engineering it also has an economic angle. It costs money to manufacture the parts of a machine. More parts cost more money, a prospect that any sane manufacturer wants to avoid. In addition, assembling a complex machine is more error-prone. Simplicity is better for building machines that work.[35]

Anybody who has struggled to understand living beings, and has despaired at their complexity, will sympathize with this yearning for simplicity. Life seems unnecessarily complex in many ways. The regulation circuit that divides insects into fourteen segments contains dozens of molecules,[36] but scientists have known for many years that just *two* molecules interacting in the right way could achieve the same goal.[37] As if to spite us, thousands of insect species segment their bodies in a way that not only took decades to understand but that no self-respecting human engineer would ever devise. And remember the road networks of metabolisms, full of redundant lanes, alternative routes, and unused back alleys. They all raise the same question: Why? Why doesn't ruthlessly efficient nature get rid of all this complexity?

The answer is "the environment"—or rather, "the environments." What looks like a wastefully complex suite of genes is actually the secret to survival *in more than one environment.*

In the kind of nutrient-poor environment in which *E. coli* has only a single carbon source to manufacture those sixty essential biomass units that include amino acids and DNA nucleotides, nearly three-quarters of the bacterium's metabolic reactions are completely dispensable.[38] Knock them out and life continues—robustness.

But environments change. If the sole source of carbon changes from glucose to ethanol, some of those "dispensable" reactions can keep the biomass factories running. Each of the eighty carbon sources from which *E. coli* can synthesize biomass needs some dedicated reactions. And carbon is only one of the essential elements—metabolizing sources of other elements needs further reactions. A large collection of metabolic reactions makes an organism viable in multiple environments. In biology, increased complexity means increased robustness to environmental change.

For the same reason of changing environments, duplicate genes often persist in the genomes of organisms. Duplicate genes, like humans, are created equal, but they do not stay that way for long. They accumulate mutations that alter their DNA and its molecular meaning, and lead to increased specialization on one environment. Some human duplicate enzymes are best at cleaving molecules in the chemical environment of the liver, whereas others operate optimally in the brain. One duplicate yeast protein is best at importing glucose into the cell when this nutrient is abundant, whereas its duplicate partner specializes at scavenging glucose when it is scarce. The redundancy of many gene duplicates is more apparent than real, because they ensure robustness to changing environments.

The world of technology provides examples as well, for even though engineers prize simplicity, they must also design for changing environments. If you want to travel with a river's current, a simple wooden raft will do the trick. But to cross the river or steer the raft, you already need more complexity: a rudder. If you want to avoid getting soaked by waves, you need a hull. To navigate upriver, you need oars or a sail. The simplest sail—the kind of square rig that Phoenicians and Egyptians had already built five thousand years ago—works well to sail downwind, but it is less effective when wind directions change, and useless for sailing upwind. To do that well you need a more complex fore-and-aft rig with two sails, a jib in front of the mast and a mainsail behind. Navigating a changing environment—current, waves, and wind—needs complex technology.

The converse is equally true, at least in biology. Over time, unchanging environments result in less complexity, because robustness becomes less important. To find examples, you don't have to look much farther than a houseplant—or, more accurately, to the insects living on one.

Aphids, also called plant lice, blackflies, and greenflies, have been blood enemies of farmers and gardeners for millennia, though only a few hundred of the more than four thousand aphid species suck the sap of agriculturally valuable plants—not only of the houseplants in suburban homes, but of cotton, fruit trees, and grain crops. They were complicit in both the Irish potato famine of the 1840s and the Great French Wine Blight of the 1850s. They are among the most destructive of all insect parasites. Yet aphids are also valuable, at least to science.[39] For deep inside them live other, even smaller organisms that can teach us a lot about both robustness and complexity.

Many people know that aphids suck sap, but fewer know that sap is a very poor food. It lacks essential molecules, including several amino acids. To get these, aphids have teamed up with relatives of the bacterium *Escherichia coli,* a species called *Buchnera aphidicola* that inhabits the bodies of aphids.

The alliance between *Buchnera* and aphids benefits both species and is also known as an endosymbiotic mutualism. It is a remarkably close relationship. *Buchnera* does not just live on or near aphids. Its cells live *inside* the aphid's cells, where they provide their host with a vital service: They manufacture essential food molecules, especially amino acids that aphids cannot manufacture themselves, and that plant sap does not contain. *Buchnera* is a tiny food factory that keeps aphids alive.

For its services *Buchnera* also gets something in return. By living inside an aphid's cell, *Buchnera* floats in a broth of molecules that is rich enough to provide almost all the food the bacterium needs. And not just food: The aphid's cell walls provide *Buchnera* with a safe and comfortable home. By dragging *Buchnera* with them everywhere, aphids insulate them from heat, cold, rain, and other environmental hazards.[40] *Buchnera* needs to take no notice of overgrazed plants, lurking predators, or any other threat. So long as its aphid home survives, the bacterium thrives. Like a vacationer idling its time away in the gently lapping waves of a tepid ocean, *Buchnera* is sheltered from a hostile world.[41]

Buchnera's vacation has been going on for a very long time. The host and the bacterium first cohabited more than a hundred million years ago, and have lived together ever since. Over that much time, one might expect significant changes to any organism. That's what happened to *Buchnera,* and those changes reveal much about the relationship between robustness and complexity.

To understand that relationship, it is useful to compare *Buchnera* to its cousin, *Escherichia coli,* that marvel of metabolic flexibility, able to survive on dozens of different food sources, and highly robust to changing chemical environments. *E. coli*'s complex metabolic network harbors more than a thousand chemical reactions, a large collection of skills for surviving in a changeable and uncertain world.

The metabolism of *Buchnera*'s ancestors once resembled that of *E. coli.* But no longer. Now its metabolic network has a puny 263 metabolic reactions. Its alliance with aphids started when dinosaurs still walked the earth, and since then it has lost nearly three-quarters of the reactions that *E. coli* still harbors. A steady stream of DNA copying errors eroded the genes that are needed for these reactions, and many of them disappeared through gene deletions that occur naturally in DNA. *Buchnera* has survived hundreds of such deletions.[42]

It takes no genius to see why *Buchnera* could survive all these deletions and become so much simpler than *E. coli.* Hundreds of genes and metabolic reactions have become superfluous in *Buchnera,* because its world has stood still for more than a hundred million years. While its host aphid struggles in an ever-changing environment, *Buchnera* bathes in a steady if monotonous diet of nutrients. To survive in a simple, unchanging world, a simple metabolism does the trick, and complexity becomes not only superfluous but wasteful.

Buchnera is special but not unique. Many microbes live on or inside other, larger organisms. Some serve their hosts, others exploit them. A well-known human example is the bacterium *Mycoplasma pneumoniae,* a cause of the "walking pneumonia" that does not tie patients to their beds. This parasite shuttles from body to human body, and relies on human cells to supply its food. Its metabolism is even simpler than that of *Buchnera:* It needs a mere 189 reactions to survive on the rich molecular buffet that human cells provide. Incredibly, it has even shed part of metabolism's universal core, the most ancient citric acid cycle. What's more, its extreme minimalism helps it resist antibiotics that attack the enzymes building the bacterial cell wall: It no longer makes this wall, and even hijacks membrane molecules from our own body to prevent its molecular innards from spilling.[43]

A corollary of this increase in simplicity is a corresponding decrease in robustness: not just robustness to mutations, but robustness to environmental change—the two are linked. A metabolism that is robust to knocking out enzyme-coding genes will also be robust to changing environments. If *E. coli* were to live in a single, fixed environment—one, for example, where glucose is the only source of carbon—it could do without 70 percent of the chemical reactions in its complex metabolism.[44] But in *Buchnera* this is no longer the case. Fully 90 percent of its 263 reactions are essential.[45] Eliminate one of them and you kill the organism.

Put another way, the metabolic road network of *E. coli* has many alternate routes, but *Buchnera*'s doesn't. It is more like a single-lane road without exits. Block it anywhere and all traffic piles up behind the roadblock—the place where an essential molecule can no longer be made. *E. coli* is robust, both to the DNA mutations that eliminate metabolic reactions and to changing environments. *Buchnera* is not robust to either.

E. coli and *B. aphidocola* are only two specks of dust in a vast library of metabolisms, and what holds for them—an organism viable in a changing environment is more complex and robust—might not be universal. We cannot examine all metabolisms, for there are too many. But we can examine many of them computationally, and thus do what pollsters do for human populations: learn about a very large population from random samples that reflect the population's properties. By exposing such a random sample of metabolisms to changing environments, we can learn whether *E. coli* and *Buchnera* are unusual or typical.

To this end, researchers in my laboratory create hundreds of metabolic networks that contain as few reactions as possible and yet can still sustain life. We call such networks *minimal metabolisms*. Make them any smaller and you destroy their ability to sustain life. We can create minimal metabolisms that can sustain life in one environment, but also in two, three, and more environments, up to dozens of them, each one differing only in available nutrients.

One lesson from such minimal metabolisms is that living in many environments *generally* requires complexity. In one study we analyzed environments that differ in their sources of sulfur, a chemical element as essential as carbon. We first identified minimal metabolisms—there are more than one—that can sustain life on only *one* sulfur source, and found that such metabolisms need fewer than twenty chemical reactions. But to sustain life on *five* different sulfur sources, a metabolism already needs at least twenty-five reactions. And to be viable in forty different environments, it needs more than sixty chemical reactions. In other words, a metabolism that can sustain life in more and more environments needs more and more reactions. It needs to be more complex.[46]

The same metabolism also becomes more robust: We can remove more and more reactions from it, while it remains viable in any one environment.[47] The more reactions a metabolism contains, the more reactions it can do without in any one environment. These reactions are neutral in one environment, but they can become essential in a different environment. Thus *E. coli* and *Buchnera* are not unusual, but special cases of a general principle: Life's complexity and robustness increase with its exposure to environmental change.[48]

With this recognition a circle is closing. Environmental change requires complexity, which begets robustness, which begets genotype networks, which enable innovations, the very kind that allow life to cope with environmental change, increase its complexity, and so on, in an ascending spiral of ever-increasing innovability. At the core of this innovability is the self-organized multidimensional fabric of genotype networks, hidden behind life's visible splendor, but creating this splendor. It is the hidden architecture of life.

CHAPTER SEVEN

From Nature to Technology

YaMoR—an acronym for Yet another Modular Robot—is an experimental robot constructed with hinged components, a bit like the segments of a millipede, that can be arranged in a straight line to undulate like a worm, or in leglike pairs to crawl like an amphibian, or even to walk like an insect. And these segments are smart, equipped with reconfigurable hardware, each segment containing a computer chip that can be rewired a bit like a human brain.[1] Built the right way, such robots could learn on the job by rewiring themselves when they need new skills.

Constructed at one of the world's leading engineering schools, the École Polytechnique Fédérale de Lausanne in Switzerland, YaMoR is a member of a growing family of modular robots created in engineering labs around the world. The family is large but its members lack much resemblance, because their body plans are very different. Some modular robots are dielike cubes, some chains of tetrahedrons. Others are clusters of balls, and still others strings of rotating wheels. Modular robots like YaMoR seem limited only by the constraints of solid geometry.

More than 540 million years before YaMoR took its first steps, an even more diverse family of body plans arose, in a burst of biological innovation known as the Cambrian explosion. It gave rise to every animal body plan now in use, along with dozens more that are extinct, like the entire phylum of limbless segmented marine animals known as the Vetulicolia. YaMoR's phylum is much more primitive than the Vetulicolia, but if the warp drives that accelerate nature's innovation could be put to work in human technology, robotic or otherwise, then the first Cambrian explosion may not have been the last.[2]

This idea—that analogs of genotype networks could accelerate innovation in technology—is not so far-fetched, as we shall see. A first hint is that innovation in nature and innovation in technology show many parallels.

Trial and error, for one thing. Thomas Edison, an archetype of inventive genius, "tested no fewer than 6,000 vegetable growths, and ransacked the world for the most suitable filament material" until he eventually stumbled upon bamboo as the best solution for the fickle filaments of his incandescent light bulb.[3] One of the dozens of quotations attributed to him sums it up: "I have not failed. I have just found ten thousand ways that do not work." The quote reminds everyone that trial and error—especially error—is as crucial to technological innovations as to biological ones. And it is no less crucial today than in Edison's time. John Backus, a co-creator of the highly successful computer programming language Fortran, which helps scientists simulate and understand the universe, from the motions of atoms to those of galaxies, said that "you need the willingness to fail all the time. You have to generate many ideas and then you have to work very hard only to discover that they don't work. And you keep doing that over and over until you find one that does work."[4]

To be sure, failure has different consequences in evolution than it does for an inventor with a failed light bulb or for a scientist with a disproved theory. A bar-headed goose with a mutant hemoglobin produced by nature's tinkering with DNA is a living experiment.[5] If the mutation improves its ability to scavenge oxygen from thin air, great. But woe unto the bird whose mutant globin can no longer bind oxygen. Its lights go out permanently.

Failure in science and technology does not usually mean bodily death, but this does not mean that ideas die easily. Sir Fred Hoyle, one of the world's most prominent astronomers and astrophysicists, went to his death in 2001 not only denying the Big Bang theory but defending the belief that flu epidemics arise when a lull in the solar wind allows extraterrestrial flu viruses to enter the atmosphere. The nineteenth century's Lord Kelvin used the laws of thermodynamics—and his Christian faith—to underestimate the age of the earth more than a hundredfold.[6] The historical battlefields of science and technology are littered with brilliant minds who took wrong beliefs to their graves. Max Planck, a father of quantum theory, observed that "a new scientific truth does not triumph by convincing its opponents and making them see the light, but rather because its opponents eventually die, and a new generation grows up that is familiar with it." Science, like nature, advances one funeral at a time.[7]

One of nature's antidotes against catastrophic failure has been accidentally co-opted by technological innovators: populations. Great inventions aren't the work of solitary geniuses any more often than nature's libraries are explored by single organisms. Despite the cliché of human innovators conjuring unimagined worlds from the depths of their minds—from Archimedes in his bathtub to Einstein in his patent office— the truth is that technological innovation depends on the same kind of crowdsourcing that biological innovation uses when its chemical libraries are explored by armies of browsers. A team of people developed Fortran, and Edison used dozens of assistants to create and test new designs for light bulbs, telephones, and telegraphs. The nineteenth-century Industrial Revolution was made possible by the rise of an entire new class—highly educated artisans far more numerous than aristocratic gentlemen scientists—who needed to make money from their work and created a flurry of inventions from steam engines to automatic looms.[8] And today, any new technology, from new cell phones to new drugs and new energy carriers, requires armies of scientists and engineers, intense competition, and myriad failures before finding success. Given the importance of trial and error, it's hard to see how it could be otherwise: The more explorers, the more solutions can be explored, with correspondingly greater odds of success.

And when the armies of technological innovators advance, they do so on many fronts simultaneously—again like nature. The American sociologist Robert Merton, who also coined the terms "role model" and "self-fulfilling prophecy," is well remembered in the history of sciences for documenting the prevalence of inventions with multiple origins—he simply called them "multiples."[9] The list is nearly endless: The relation between the heat and pressure of a gas is known as Boyle's law in English-speaking countries, and as Mariotte's law to Francophones, since the same phenomenon was independently derived by Robert Boyle and Edme Mariotte. Robert Fulton, the Marquis de Jouffroy, and James Rumsey are all "inventors" of the first steamboats. The thermometer has at least six different inventors, and calculus, as is well known, was created almost simultaneously by both Isaac Newton and Gottfried Wilhelm Leibniz.[10] Elisha Gray filed for a patent on a working telephone the same *day* as Alexander Graham Bell (though Bell won the subsequent legal wrangle over priority).[11]

Multiple origins are possible because the problems of technology—like those of biology—have multiple solutions. Among the best-documented biological examples are the innovations that remove carbon dioxide from the atmosphere, a process called carbon fixation by biologists and carbon scrubbing by engineers. The premier biological carbon scrubber is the one used by plants, an enzyme that attaches carbon dioxide to a sugar called ribulose-1,5-bisphosphate. It then modifies this carrier further, such that the carbon dioxide eventually becomes part of the plant's body. This is not only how plants grow and how carbon gets into the fossil fuels we burn. It is also the way carbon dioxide is fed back into the cycle of life. But plants are not the only organisms that fix carbon: Some microbes attach it to the carrier molecule acetyl-CoA, and others add it to molecules from the ancient citric acid cycle.[12] Environmental engineers—they seek to solve the same problem to forestall catastrophic climate change—have already come up with several further carbon-scrubbing technologies that use molecules like monoethanolamine and sodium hydroxide.

Other examples of Merton's multiples are legion. An automobile's engine can use reciprocating pistons or an eccentric rotor, and it can trigger fuel combustion with the spark plug of a gasoline engine or through the heat of compression in a diesel engine. Organisms can detect light waves using a flexible single lens, or the rigid compound eye of a fly.[13] Antifreeze proteins of Arctic and Antarctic fish, transparent crystallins that originate from different enzymes, and highly diverse oxygen-binding globins are all multiple solutions to similar problems.

Another commonality of both technological and biological innovations is that they endow old things with new life. The history of technological innovation is practically saturated in examples. In the words of the writer Stephen Johnson, Johannes Gutenberg borrowed "a machine designed to get people drunk"—a screw-driven wine press—"and turned it into an engine of mass communication."[14] Microwave ovens heat food with a technology originally developed for radar—a radar engineer discovered its heating powers when it melted a chocolate bar in his pocket.

(The first commercial version was called the "Radarange.") The light-weight synthetic Kevlar, originally developed to replace steel in racing tires, has been co-opted for bulletproof vests and steel helmets. The same principle is at work even in mundane contraptions that barely deserve the label "innovation." A door placed on two sawhorses can make a great desk. Boots can serve as low-tech doorstops. Milk crates can make wonderful filing cabinets, and so on.[15] Edison said it well: "To invent, you need a good imagination and a pile of junk."[16]

In 1982, the paleontologists Stephen Jay Gould and Elizabeth Vrba baptized the biological version of this phenomenon "exaptation."[17] (Once again, Darwin had gotten there first—he reminded the *Origin*'s reader that "an organ originally constructed for one purpose . . . may be converted into one for a widely different purpose.")[18] A classic example of exaptation is the bird's feather, a complicated assemblage of fibrous proteins called keratins, the same proteins that make up the scales of reptiles. The first feathers were most likely involved in insulation or waterproofing, and were only later co-opted—"exapted"—for flying.[19] Such exaptations also abound in molecules, including those regulators that help make feathers. One of them is a protein called "Sonic hedgehog"—yes, it is named for the character in the eponymous video game—that helps control the growth of both fingers and spinal cord in your body, but was co-opted in birds to sculpt feathers.[20] Likewise, a regulator protein that helps shape legs in some organisms was co-opted to paint eyespots in others, and metabolic enzymes were co-opted as the crystallins of lenses.

Such co-option is a special case of a final parallel between nature and technology: Innovation is combinatorial. It combines old things to make the new. We encountered combinatorial innovations first in metabolism, where new combinations of old reactions turned toxins like pentachlorophenol into fuels to manufacture biomass, and allowed our ancestors to detox their bodies by synthesizing urea. In proteins they are new combinations of old amino acids, and in regulation circuits they are new combinations of interacting regulators. But a technological innovation like the aviation-transforming jet engine is just as combinatorial.[21] Its three principal parts are a compressor that increases the air's pressure, a combustion chamber that mixes air with fuel and ignites the mixture, and a rotating turbine that generates thrust from the expanding mixture. None of these three elements were new when half a dozen British, German, Hungarian, and Italian engineers—remember those multiples— were building the first jet engines in the twentieth century. The earliest compressors were bellows used to run forges more than two thousand years ago, and have been used in industrial blowers ever since. Combustion chambers had been central to both steam locomotives and internal combustion engines. Archimedes had invented a screw turbine in the third century BC, and the first gas turbine was patented in England in the late eighteenth century.

The jet engine is scarcely unique as a combinatorial innovation. Decades ago, social scientists like the economist Joseph Schumpeter and the sociologist S. Colum Gilfillan argued that combining the old to make the new is essential for technological innovation. In his book *The Nature of Technology* the economist W. Brian Arthur goes further to say that even entire "technologies somehow must come into being as fresh combinations of what already exists."[22] So too in biology, as we heard in previous chapters: *All* evolutionary innovations, discovered as they are in searches through nearly infinite libraries, are combinatorial, just as new books combine old letters into new meanings.

Trial and error. Populations. Multiple origination. Combination. With all these parallels between technology and nature, it is little surprise that technologists would try to mimic nature's innovability. And I do not just mean biotechnologsts, whose innovations are already legion, from the enzymes in our laundry detergents that turn my ten-year-old's mud-caked pants spotless, to the engineered forms of insulin used by diabetics, to crops that have been genetically modified to produce a powerful bacterial toxin that kills insects preying on them. Because biotechnology uses biological materials, it already takes advantage of nature's libraries. The bigger question is whether technologies built on man-made materials—glass, plastic, or silicon, like YaMoR—can do the same.

Technologists made a big first step to answer this question when they realized that evolution follows an *algorithm,* a recipe so simple and stereotypical that it could be executed by a machine.[23] By altering DNA, mutations create organisms with new phenotypes, and selection allows some of them to survive and reproduce. Mutate. Select. Repeat over and over.[24] Those technologists who know their automata—computer scientists—created an entirely new research field from this insight, one that revolves around *evolutionary algorithms:* recipes that use some form of mutation and selection to solve really difficult real-world problems, entirely inside a computer.[25]

An especially famous and difficult such problem is known as the traveling salesman problem, a mathematical puzzle first formulated in the mid-nineteenth century by the Irish mathematician William Rowan Hamilton. The basics are simple: A salesman has to visit dozens of potential faraway clients all on a regular basis. Each of them lives in a different city. The salesman spends a lot of time on the road and in the air. To spend more time with his family, he would like to keep each trip as short as possible and still visit all cities. The problem is to find a travel route that takes him to each city in turn, and then back home at the end of the trip, as quickly and efficiently as possible.

This problem is a lot harder than it sounds. For a mere handful of cities, anybody could design the shortest possible route. But once the number of cities increases beyond a few dozen, the optimal route becomes surprisingly difficult to find. The traveling salesman problem is what computer scientists call a "nondeterministic, polynomial-hard" problem.[26] It is among the hardest of all puzzles that can be imagined, mainly because the number of potential solutions increases exponentially as new cities are added.

Thousands of papers have been written about this problem, because it isn't just for sales forces. Designers of computer chips face it too. Such chips contain thousands to millions of components that are wired together and exchange data. Because short wires between them can save energy and accelerate computations, designers of such chips need to find shortest possible routes that connect many circuit components ("cities"). Truckers who deliver goods to multiple department stores also know this problem, as does Federal Express, and so do school buses that pick up multiple children within a school district. Even bumblebees face it: A foraging bee might visit hundreds of flowers before returning to the hive, and cannot afford to waste energy on excessive travel.[27]

It's possible to calculate *perfect* solutions to the traveling salesman problem for up to thousands of "cities," using sophisticated mathematical techniques with deceptively simple names like "cutting plane" and "branch-and-bound." Such techniques can produce *near-perfect* solutions for up to millions of cities. But even an algorithm as blind and mindless as that used by biological evolution can tackle the problem, by instructing a computer to start with a completely arbitrary solution—any solution at all, no matter how bad. The computer's program then mutates the solution randomly, changes the route between a few cities (or stores, or schools, or flowers), and checks to see whether the new route is shorter than the old one. If so, the program selects that mutant. Next, it tries a new mutation, and checks whether this mutation shortens the route. If not, it rejects the mutation, reverts to the original route, and begins again with a new mutation. Over many generations, this simple algorithm produces shorter and shorter routes, and eventually arrives at good if not perfect solutions.[28]

Evolutionary algorithms like this are also being put to work in some surprising places. Military planners use them to plot optimal routes for unmanned drones that patrol hostile territories.[29] Cryptographers use them to encode sensitive data. Fund managers use them to predict the movements of financial markets. And automotive engineers use them to change how an engine operates, by optimizing the time or pressure at which fuel is injected into the engine. And those algorithms do indeed improve fuel efficiency.

What they don't do is revolutionize engine design.[30]

The evolutionary algorithms that mimic biological evolution are powerful tools, but they're still missing something. They are still deficient in the recombination department so central to biological innovations.[31] Nature is better at recombination, much better, for one simple reason: standards.

As we've seen in chapter 2, standards like the universal energy standard ATP and the universal genetic code are a sign of life's common origin. The absence of such standards makes recombination more difficult in technology, which often uses ingenuity as a substitute. It took plenty of ingenuity to recombine a compressor, a combustion chamber, and a turbine to lift a plane's many tons of metal into the air. The same holds for the combustion engines of today's cars, whose parts are connected through one-of-a-kind links, like pistons and valves that need to be precision engineered to fit a cylinder. And likewise for signature inventions of the Industrial Revolution. The first practical steam engines combined steam-powered toys originally invented in second-century Alexandria with vacuum pumps from seventeenth-century Germany. Bench vises joined two of the simple machines of antiquity, a lever-handle and a screw. The first bicycles combined three, the wheel, the lever, and the pulley. All of these combinatorial innovations required ingenuity.

It's not that standards are absent from technology, far from it. Technology relies not only on universal laws of nature established by science, but also on standardized ways of measuring quantities like temperature, mass, or electric charge. But most technologies are deficient in a certain kind of standard, the one that allows nature to combine the old to make the new. Nature needs these standards, precisely because it does not have the ingenuity of human inventors.

All the different things that proteins can do—catalyze reactions, transport molecules, support cells—emerge from strings of building blocks connected in the same way, through a standardized chemical connection called the *peptide bond,* where an atom of nitrogen from one amino acid bonds with an atom of carbon from its neighbor. Even though each amino acid has a different shape, they can all connect in the same way, because they share a universal interface. And this standard, used by all organisms, has made life as we know it possible. It allows nature to cobble together—blindly, without any ingenuity—the astronomical numbers of genotypes needed to find innovations.

Standards that make recombination mindlessly easy do not just exist in proteins. RNA strings also use a standardized chemical bond to link their parts. Furthermore, life's standard of information storage—DNA—allows bacteria to exchange genes and create new metabolisms from new combinations of the enzymes they encode. And finally, regulation circuits use a standardized way to regulate genes, based on the principle that regulator proteins can bind specific short words on DNA, allowing nature to combine old regulators into countless new circuits by changing these words.[32] If we could take a small number of different objects, create a standard way to link them, and recombine them into every conceivable configuration—mindlessly—our powers to innovate could be just as immeasurable as those of nature.

Such standardization is clearly not beyond human technology: Our hardworking Lego blocks hint at it, and so does a much older human technology.

The sixteenth-century Venetian Andrea Palladio may be the most influential architect in Western history. Throughout a long and success-ful career, he conceived at least sixteen of the urban palazzos that housed Venice's wealthiest families, built thirty country villas, and designed sev-eral churches.[33] The floor plans of Palladio's buildings are not identical. Far from it. They differ in a thousand ways. His buildings are different in size, shape, orientation, and the arrangement of their rooms. Yet they share an architectural essence, even if most of us would find that essence hard to pinpoint. More than twenty years ago, the art historian George Hersey and computer specialist Richard Freedman searched for this ele-ment, the secret behind Palladian floor plans.[34] If rules behind these plans exist, they thought, then we must be able to find an algorithm to create any number of Palladian floor plans.

To distill the essence of Palladio's style, they analyzed dozens of Pal-ladian villas: how their rooms were oriented, how their walls were placed, whether the lengths of adjacent rooms had certain proportions, and so on. And they succeeded. Their work culminated in a computer program that executes a recipe to create new Palladian floor plans. Its creations differ in many details, in the sizes, orientation, and arrangements of rooms, but all of the new designs are recognizably Palladian.[35]

The algorithm behind this program starts from the outline of a building, usually rectangular in shape, and draws vertical or horizontal lines through it—walls that subdivide the building into rooms. One such line would split the building into two rooms, two parallel lines would split the building into three rooms, and so on. Each room could again be split by horizontal or vertical lines, and the resulting rooms can be split further, and so on, until only rooms with a desired size have been cre-ated. By splitting a rectangle multiple times, each time either vertically or horizontally, each time by one or more walls, one can create an infinite variety of floor plans.

The rules underlying Palladian floor plans are neither arbitrary nor random. Nor are they complicated. For example, when a room is split in two with one line, it is usually split either in the middle or such that one of the resulting rooms is twice as long as the other. If it is split with two parallel lines, most splits create three rooms where the middle room is twice as long as the others. By combining these and a few other rules, this most celebrated style of Renaissance architecture can be re-created by a computer.[36]

Nature's stereotypical recombination doesn't work *exactly* the same way. Small amino acid parts are combined to make new proteins, while the outline of a Palladian building is subdivided to create a floor plan. But the similarities are more important: Both use a small number of standard building blocks and an even smaller number of rules to create enormously complex objects. And if these similarities exist already in preindustrial architecture, the odds that they exist in postindustrial technology—unappreciated perhaps—should be even better. [37]

As in the technology behind robots like YaMoR: digital electronics.

Each of the myriad transistors on an integrated circuit—the kind of computer chip guiding a robot—is really nothing more than a tiny elec-tronic on-off switch that responds to electronic impulses: one denoted as a "0" turns the switch off, whereas the other, a "1," turns it on. Together, these transistors transform an arriving stream of data into a departing stream that also contains only zeroes and ones, the familiar *bits* or binary digits from the simple two-letter alphabet that computers read. Mathe-maticians describe this process more precisely, by saying that a circuit calculates the values of a *function,* that it takes an *input* and computes an *output.*[38] The kinds of functions that a digital circuit computes are named after the English mathematician and philosopher George Boole, who wrote about them in his 1854 book, *An Investigation of the Laws of Thought on Which Are Founded the Mathematical Theories of Logic and Probabil-ities.* Boole's new branch of mathematics was an enormous advance, and Boolean logic functions, as they are called today, are still at the heart of all modern computers.

One of the simplest such functions is the AND function, which is needed whenever one searches, for example, an electronic database of sheet music for a specific composition, such as Mozart's *Magic Flute*. The search engine would search for compositions containing the word "Mozart," and it would report a yes or a no—encoded by the bits 1 and 0—for all compositions containing this word. It would do the same for the word "magic," yielding two possible answers for each of these two partial questions, which makes four possible combinations of answers. Their digital one-zero version can be written in a form that mathematicians use to describe Boolean logic functions, a *truth table*. In the truth table of the AND function shown in figure 20a, the function's input, the four possible combinations of partial answers are written to the left of the vertical line, and to the right are the possible final answers—the output—again encoded as zeroes and ones. Merely one of the four lines

a)

Mozart?	Magic?	Mozart AND Magic?
1	1	1 (yes)
1	0	0 (no)
0	1	0 (no)
0	0	0 (no)

b)

Mozart?	Magic?	Mozart OR Magic?
1	1	1 (yes)
1	0	1 (yes)
0	1	1 (yes)
0	0	0 (no)

c)

Mozart?	NOT Mozart?
1	0 (no)
0	1 (yes)

FIGURE 20. Truth tables

contains a yes as the final answer—only if both partial answers are yes has a score for *The Magic Flute* been found.

If you wanted to find compositions whose title contained the word "Mozart" *or* "magic" (*or* both), the search engine would have to compute another Boolean function: the OR function. Same input stream—everything with the word "Mozart" and "magic"—but a different rule. In this case, the output is a yes if at least one partial answer is a yes (figure 20b). As a result, the OR function would report not only *The Magic Flute* but also every other one of Mozart's 626 pieces,[39] as well as Santana's "Black Magic Woman," Stevie Wonder's "If It's Magic," and hundreds more. An even simpler Boolean function, the NOT function (figure 20c) turns a yes into a no, and could help you find all compositions without the word "Mozart" in them.

These and other more exotic Boolean logic functions—XOR, XNOR, NAND, NOR—allow us to translate complex questions from natural languages into the strings of binary numbers that rule the world of computers. What's more, binary numbers can encode any decimal number, and can be added, subtracted, multiplied, and divided just like decimal numbers. The integrated circuits of even the most sophisticated computers perform nothing but basic arithmetic and simpler Boolean logic functions like the AND function. With the simplest of all possible alphabets—zeroes and ones—and Boolean logic, digital computers can recognize images, encrypt information, deliver voice mail, and predict next Tuesday's weather. Arithmetic, it turns out, is even more important than we learned in grade school.

Another remarkable fact about Boolean functions is that even the most complicated Boolean function can be computed by stringing together simpler functions, such that the output of one function becomes the input of another. It's a bit like multiplication (3 × 4), which can be written as a series of additions (4 + 4 + 4). More than that, even though the number of possible Boolean functions is virtually infinite, each of

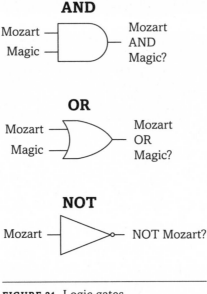

FIGURE 21. Logic gates

them can be computed by stringing together only AND, OR, and NOT functions. This matters for computers, because in an integrated circuit, transistors are wired into units of computation that compute a simple Boolean function and that are known as *logic gates*. Figure 21 shows some of the symbols that chip designers use for the simple AND, OR, and NOT gates. Each gate has one or two wires to the left for its input bits, and one on the right for its output bit. In figure 22 a few gates are wired to perform the simplest possible arithmetic operation of adding two binary digits—this already requires six logic gates, each with multiple transistors.[40] Today's chips of course add, subtract, multiply, and divide much longer numbers of sixty-four binary digits, and require millions of gates.[41]

Most integrated circuits are hardwired in the factory, but robots like YaMoR are equipped with programmable hardware, chips that can be rewired to alter what some logic gates do—changing an AND gate into

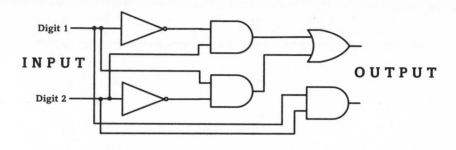

FIGURE 22. A circuit adding two binary digits

an OR gate, for example—and how these gates are wired. Some programmable chips can even be rewired while the chip is working.[42] With more
than a million logic gates, such chips are not just simple toys but powerful and flexible computing engines that could eventually help machines
learn much as we do—by rewiring their own hardware—and create
autonomous robots that can not just explore the world but also learn
about its potholes and other pitfalls.[43]

If this sounds familiar, it's because such learning resembles evolution, which alters life's genotypes one molecule at a time. A programmable circuit's logic gates and wiring are an analog of a genotype that can
be altered to explore new computations, the analog of new phenotypes.
Like evolution, the learning process requires plenty of trial and error. It
reinforces good behavior, and punishes bad behavior. (But not as severely
as evolution does. If your—or a future robot's—golf game is weak, you
may need to improve your stance, your grip, or your swing, but you need
not die.) What is more, such learning need not destroy old knowledge. As
you learn to play golf, you can still sit, walk, run, or jump, even though
the neural circuit responsible for these skills gets rewired. And the parallels to evolution don't end here. The wires connecting logic gates are
generic links, just as flexible as the peptide bonds of proteins, because the
output of any one gate can be fed into any other gate.[44] As with proteins,

such links can be built, destroyed, and modified without much ingenuity to create electronic circuits that can be wired in astronomically many ways.[45]

Standard links. Few kinds of logic gates. These principles already suffice to create chips that play chess more powerfully than mankind's best, to find a single page in a million different books, or to "print" objects in three dimensions. The capabilities of real-world programmable circuits are so reminiscent of nature's innovation powers that they suggest a profound question: Are entire libraries of digital circuits—huge circuit collections that can be created through recombining logic gates in every possible way—organized like the library of biological circuits? The answer can tell us whether the warp drives of biological innovation could be mounted on the spaceships of technological innovation.

Karthik Raman provided this answer. A graduate from the Indian Institute of Science, one of India's top universities, Karthik came to my lab as a postdoctoral researcher. And he did not come alone. He also brought an effervescent enthusiasm for science, dogged tenacity in the face of failure—as inevitable in science as in evolution—and a wizardlike talent for analyzing complex data. When I invited him to map libraries of programmable circuits, he jumped right on it.

Although commercially available programmable chips have more than a million gates, some back-of-the envelope calculations convinced us that we should study smaller circuits. A library of circuits with a mere sixteen logic gates contains 10^{46} such circuits—a number already large beyond imagination—and this number increases exponentially with the number of gates, to 10^{100} circuits with only thirty-six gates.[46] Huge numbers like this also made the decision of whether to build circuits in hardware or to study them in the computer easy: Millions of circuits are most easily analyzed inside a computer.[47]

A sixteen-gate circuit could in principle compute 10^{19}—a million trillion—Boolean functions, but we didn't know whether the library's circuits encoded that many.[48] Perhaps its circuits could compute only a few functions, like addition or multiplication. To find out, Karthik first cast a net wide enough to haul in as many volumes as possible from the circuit library. He created many circuits with random wirings, two million of them, and found that they computed more than 1.5 million logic functions, only a few of them as familiar as the AND function. Even though he had hauled in only a small fraction of circuits—there were still 10^{40} circuits left, and 10^{12} times more functions to explore—his enormous catch taught us that even simple circuits can compute numerous Boolean functions.

Because the library hosts many more circuits than there are functions—10^{26} times more, to be precise—we knew that there must be multiple synonymous texts, circuits computing the same logic function, but we didn't know how they were organized. To find out, Karthik started with a circuit that computed an arbitrary logic function and changed it to one of its neighbors in the library, for example by reconnecting the input of one gate to the input of another. If this "mutated" circuit still computed the same logic function, Karthik kept it. If not, he tried another rewiring, and repeated that until he had found a circuit with the same function. From that new circuit he took another step, and another, and so on, such that each step preserved the circuit's function. Karthik started random walks like this from more than a thousand different circuits, each one computing a different function that needed to be preserved.

The networks of circuits he found reached even farther through the library than the genotype networks from earlier chapters: From most circuits one could walk *all the way* through the circuit library without changing a circuit's logic function. Two circuits may share nothing, not a single gate or wire, except the logic function they express, yet they can still be part of a huge network of circuits connectable through many small wiring changes. What's more, we found that this holds for every single function we studied. It is a fundamental property of digital logic circuits.[49] The library of digital electronics is like biology, only more so.

Karthik next turned to the neighborhoods of different circuits computing the same function, created all their neighbors, and listed the logic functions that each of them computed. He found that these neighborhoods are just as diverse as those in biology. More than 80 percent of functions are found near one circuit but not the other.[50] This is good news for the same reason as in biology: One can explore ever more logic functions while rewiring a circuit without changing its capabilities. A circuit's neighborhood contains circuits with some sixty new functions, but a mere ten rewiring steps make a hundred new functions accessible, a hundred rewirings put four hundred new functions within reach, and a thousand changes can access almost two thousand new functions.[51]

And the similarities continued. Earlier I mentioned the multidimensional fabric of biological innovability, the almost unimaginably complex, densely woven tissue of genotype networks. Karthik found that it has a counterpart in the circuit library, where a circuit with any function could be reached from any starting circuit by changing only a small percentage of wires. A fabric just like that of life's innovability exists in digital electronics, and it can accelerate the search for a circuit best suited for any one task.[52]

Circuit networks thus have all it takes to become the warp drives of programmable hardware, in precisely the same way that genotype networks are the warp drives of evolution. They have the potential to help future generations of YaMoRs learn many new skills, from simple self-preservation like avoiding deadly staircases to complex skills like doing the dishes or playing ball with children. In this vision, their digital brains can rewire themselves step by little step, and explore many new behaviors, while being able to preserve old behavior—conserving the old while exploring the new.[53] I wouldn't even be surprised if our brains used a similar strategy to learn. We already know that they continually rewire the synaptic connections between our neurons, but perhaps our brains also explore new connections in the same way that organisms explore a genotype network. If so, the very same principle allowing biological innovation could be at work in the engines of human creativity.

Unfortunately, our ignorance in this area is still nearly absolute. We know next to nothing about the material basis of human creativity. We *do* know, however, that the kind of creativity we discovered is not free, because Karthik found its price tag. It was a familiar one.

When Karthik analyzed logic circuits that differed in their complexity—their number of logic gates—he found that the simplest circuits could not be rewired without destroying their function. Change one wire in such a circuit, and you destroy the circuit's function. Every gate and every wire matters. Such simple circuits have no innovability, because they cannot explore new configurations and computations. For rewiring, one needs more complex circuits. The more complex they are, the more rewiring they tolerate. Their apparently superfluous gates and wires are like collections of spare parts—piles of Edison's precious junk— that help compute new digital functions. Just as in biology, innovability comes from complexity, apparently unnecessary, but actually vital. This is one of nature's lessons for innovable technologies: If we want to open nature's black box of innovation, Ockham's razor is much too dull. Like oil and water, simplicity and innovability don't mix.

This doesn't mean that simplicity and elegance are absent from powerful innovable technologies. Quite the opposite. But they hide beneath the visible world. The basic principle behind them is simplicity itself: With a limited number of building blocks connected in a limited number of ways, you can create an entire world. Out of such building blocks and standard links between them, nature has created a world of proteins, regulation circuits, and metabolisms that sustains life, that has brought forth simple viruses and complex humans, and ultimately, our culture and technology, from the *Iliad* to the iPad. The simplicity and the elegance of innovable technologies are hidden behind the visible world, just like nature's libraries, whose faint reflection we see in the Tree of Life, like a shadow in Plato's cave.

EPILOGUE
Plato's Cave

I n October 1970, the magazine *Scientific American* published a description of the Game of Life, a creation of the British mathematician John Conway, and a simplification of ideas on building self-replicating machines proposed by the polymath John von Neumann. Not requiring a human player, the "game" can unfold inside a computer on a two-dimensional grid of cells, each of which can be either "on" (alive) or "off" (dead). Each square on Conway's grid has eight neighbors, and a very simple set of rules determines their status. For example, if a cell has fewer than two live neighbors it turns off. In the game's lingo it "dies." Same result if it has between four and eight live neighbors. However, if the cell has two or three "live" neighbors, it gets to live. The final rule: A dead cell with three live neighbors is reanimated.

And that's it. But depending on which cells are on and off when the game starts, what follows is anything but simple. Enormously complex patterns can emerge, a huge and unpredictable variety of forms, including "self-replicating" clusters of cells that spawn more of themselves. And from these simple beginnings, the Game of Life can go on indefinitely, creating complex patterns that never repeat or terminate.

Like life itself.

The game is a metaphor rather than a model for life, but it reflects a broader human aspiration: to understand life and its diversity through the language of mathematics and computation. This aspiration is much older than the game. Seventeen years after the publication of the *Origin*, Charles Darwin wrote in his autobiography, "I have deeply regretted that I did not proceed far enough at least to understand something of the great leading principles of mathematics, for men thus endowed seem to have an extra sense."[1]

Like the zoopraxiscope that filmed Sallie Gardner four years before Darwin's death, Darwin's work sparked a revolution, but even if he had been a mathematician he would have been in the dark about the hidden architecture of life—he could not even know that it existed. To illuminate nature's giant libraries, the flames of this revolution would require more fuel than Darwin's theory.

Biology and mathematics first needed to become fully intertwined, which would take another century. It began with the mathematics of Sewall Wright and R. A. Fisher, which bridged the gap between traditional Darwinism and Mendelian genetics, and led to the modern synthesis that allowed the first accurate predictions of how fast natural selection can help innovations spread. Another half century had to elapse until systems biology taught us how molecules cooperate to produce the complex behavior and phenotypes of life. In doing so, it showed us that cells are vastly more complex than the simple elements of the Game of Life. Through regulatory circuits operating a bit like the neural networks of our brains, they perform sophisticated computations that regulate their own molecules and help them survive. And while these circuits are very different from digital computers—for one thing, they are self-assembled from organic molecules— they hint at a deep unity between the material world of biology and the conceptual world of mathematics and computation, a unity that Conway and Darwin could barely have guessed at.

The mathematical perspective of systems biology also allowed us to decipher the staggeringly complex phenotypic meaning of genotypic texts in nature's libraries, which is crucial to understanding innovability. It led us to identify genotype networks, and to grasp that genotype networks are the common origin of the different *kinds* of innovations—in metabolism, regulation, and macromolecules—that created life as we know it. They propelled life from its very beginnings to single-celled organisms, from our bacterial and eukaryotic ancestors to primitive wormlike creatures, fish, amphibians, mammals, and all the way to humans, spanning billions of generations.

More than that, the mathematics of biology allowed us to see that these libraries self-organize with a simple principle, as simple as the gravitation that helps mold diffuse matter into enormous galaxies. This principle—that organisms are robust, a consequence of the complexity that helps them survive in a changing world—brings forth the intricate organization of these vast libraries.

These libraries and their texts differ fundamentally from the muscles, nerves, and connective tissues that an anatomist dissects and that we can touch with our bare hands. They are not even like cellular organelles visible through a microscope, or the structure of DNA revealed by X-ray crystallography. They are concepts, mathematical concepts, touchable only by the mind's eye.

Does that mean they exist only in our imagination? Did we discover them or invent them?

The question whether knowledge—especially mathematical knowledge—is created or discovered has occupied philosophers for more than twenty-five hundred years, at least since Pythagoras and certainly since Plato. Plato saw our visible world as a faint shadow cast by the light of a higher, timeless reality on the poorly lit walls of a cave we inhabit. Platonists posit that we *discover* truths, which come to us from a higher reality. They exist even if nobody is there to see them, like the dark side of the moon. Others, such as the Austrian philosopher Ludwig Wittgenstein, argue that mathematical truths are *invented*—in Wittgenstein's words, "the mathematician is an inventor, not a discoverer."[2]

Platonism has the upper hand in this debate, even though Plato himself was unaware of the best argument for it. It is the startling congruence between mathematical theorems and physical reality, encapsulated in a dictum often attributed to Galileo Galilei: "Mathematics is the language in which God wrote the universe." (Words that should give any naïve creationist pause.) The Hungarian-born Nobel Prize–winning physicist and mathematician Eugene Wigner called it "the unreasonable effectiveness of mathematics in the natural sciences."[3]

Unreasonable indeed: We know no reason why Newton's laws should predict so much more than the speed of a falling apple, phenomena as different as the rotation of planets and the shaping of galaxies. Except they do. And so do countless other mathematical laws that explain phenomena so remote in space and time that we will never experience them directly. The nexus between math and reality is so tight, in fact, that the Swedish theoretical physicist Max Tegmark argues that the entire universe *is* mathematics.[4]

But the "unreasonable effectiveness" of math is not the only reason to believe in the reality of nature's libraries and their genotype networks. Another is that the technology of the twenty-first century grants us unrestricted access to these libraries. In so doing it can shift the debate about discovery versus invention—uncomfortably abstract for millennia—from its traditional focus on languages like that of mathematics to incorporate experimental science. The reason is that we now can read individual volumes in nature's libraries. We can, for example, manufacture any volume of the protein library—*any* amino acid sequence at all—and study its chemical meaning with the instruments of biochemistry. Many of these volumes were discovered by other organisms long before us, and their molecular meaning has surprised us greatly, as antifreeze proteins, crystallins, and Hox regulators testify. It's a safe bet that nature's libraries will continue to surprise us—more than anything we just invented.

When we begin to study nature's libraries we aren't just investigating life's innovability or that of technology. We are shedding new light on one of the most durable and fascinating subjects in all of philosophy. And we learn that life's creativity draws from a source that is older than life, and perhaps older than time.

ACKNOWLEDGMENTS

Some key collaborators are mentioned by name in the text, but I am greatly indebted to numerous others, from graduate students to postdoctoral fellows and faculty colleagues at multiple universities. I am especially grateful for discussions with research associates in my laboratory, among them Aditya Barve, Sinisa Bratulic, Joshua Payne, José Aguilar-Rodríguez, and Kathleen Sprouffske. In many conversations superficially unrelated to this book, they have unknowingly sharpened my thinking about the material represented herein. My thanks also go to the numerous colleagues and fellow visitors whom I have encountered over the years at the Santa Fe Institute, which has remained a wellspring of new ideas and stimulation. Special thanks go to Jerry Sabloff and Doug Erwin, who provided feedback on an early draft of the manuscript. I am also indebted to Cormac McCarthy, who not only read this early draft but also provided many useful editorial comments. (Bowing to his avowed aversion to punctuation, this book is free of semicolons.) My faculty colleagues at the University of Zürich deserve thanks for helping create the kind of research environment in which projects like this can thrive.

ACKNOWLEDGMENTS

Bill Rosen taught me that a good editor can turn caterpillars into butterflies. His guidance was instrumental at all stages of this project. He did an outstanding job and I cannot thank him enough. He and my agent, Lisa Adams, also helped me navigate the treacherous waters of the publishing industry. Lisa superbly handled all contractual matters. Furthermore, I am indebted to Niki Papadopoulos of Current for editorial support. Her incisive comments and questions have helped improve the manuscript greatly. She and her assistants, Kary Perez and Natalie Horbachevsky, have also promptly and patiently handled numerous queries. Last but not least, thanks go to my family for their benevolent tolerance of my moods when the roller coaster of the writing process went on one of its downturns.

NOTES

PROLOGUE: WORLD ENOUGH, AND TIME

1. The assumption that these processes proceeded through most of the earth's history at the same rate as today is a principle of geology known as uniformitarianism.
2. See Zimmer (2001), 60.
3. See Burchfield (1974). A relevant passage can be found in chapter 10, page 338, of the sixth edition of Darwin's central opus, *On the Origin of Species by Means of Natural Selection*. See Darwin (1872). Darwin published six English-language editions of this book in his lifetime, each different from the one before. In these endnotes I cite page numbers from the sixth edition, as reprinted by A. L. Burt (New York), but usually the first edition, i.e., Darwin (1859), is cited.
4. See Burchfield (1990), 164. This often-quoted anecdote obscures the true reason for Kelvin's error (which is unimportant for my point here). It was his assumption that thermal conductivity is uniform throughout the earth's interior, as discussed in England, Molnar, and Richter (2007).
5. See Sibley (2001).
6. See Schwab (2012), 188, as well as Tucker (2000).
7. See Goldsmith (2006).
8. Human nails and birds' claws are composed of proteins from different keratin subfamilies, known as α-keratins and β-keratins respectively. See Greenwold and Sawyer (2011) for the origin of β-keratins.
9. See Kappe et al. (2010).
10. See Shimeld et al. (2005) and Feuda et al. (2012). Vertebrates themselves originated in the Cambrian explosion more than five hundred million years ago, but some of their proteins may be much older.

11. This number has been estimated as being no greater than 10^{90}. See "Observable universe," Wikipedia, http://en.wikipedia.org/wiki/Observable_universe, for a more conservative, smaller estimate.

12. A year has 365 days, and if the universe is on the order of 2×10^{10} years old, winning a jackpot every day adds up to only 7.3×10^{12} jackpots, a ridiculously small number compared to what is needed.

CHAPTER ONE: WHAT DARWIN DIDN'T KNOW

1. A thorough and well-sourced account of the history of biology through the mid-twentieth century is Mayr (1982). I will cite extensively from it.

2. See Mayr (1982), 362.

3. Ibid., 390.

4. Ibid., 351.

5. Ibid., 363.

6. Ibid., 259.

7. See Whitehead (1978), 39.

8. This essentialist concept of a species is also sometimes called the typological species concept. See Futuyma (1998), 448.

9. Successful hybridization that creates new species is not uncommon, especially in plants. See Futuyma (1998). Because bacteria do not reproduce sexually like we do, the concept of a biological species does not apply to them. Nonetheless, they frequently exchange genetic material through a process known as lateral gene transfer, and are thus even more plastic than species of higher organisms. See Bushman (2002).

10. See Mayr (1982), 304.

11. I note that *Eupodophis* itself was only discovered recently. See Houssaye et al. (2011).

12. See Gilbert (2003). Darwin himself made contributions to this area through his extensive studies on barnacles (Cirripedia).

13. See Mayr (1982), 439.

14. Aside from his contemporary Alfred Russel Wallace, who proposed a similar theory at about the same time, Darwin is peerless in his radical application of the concept of natural selection, and in the body of evidence he accumulated for its importance. The concept of natural selection existed long before Darwin's theory, but selection was then usually thought to help eliminate degenerate forms, not to help gradually improve existing forms. See Mayr (1982), 488–500.

15. See Darwin (1872), chapter 1, page 12.

16. See Mayr (1982), 710.

17. Mendel's laws are summarized in biology textbooks, such as Griffiths et al. (2004). In some Mendelian traits, the offspring of two pure-breeding parents can be intermediate between the parents. The particulate nature of genes can

then still be revealed in the second generation, where some individuals display the parental phenotype.

18. Mendel (1866).

19. See Kottler (1979), Corcos and Monaghan (1985), Mayr (1982), 728, as well as Schwartz (1999), chapter 7.

20. See Johannsen (1913). The term *pangene* comes from the word *pangenesis,* the ancient notion that all parts of a body, including eyes, hair, and nails, contribute to inheritance. According to pangenesis, brown-eyed parents, for example, tend to have brown-eyed children because eyes contribute to whatever material a man and woman exchange in procreating. Darwin also believed in pangenesis. See Mayr (1982), 693. We now know pangenesis to be wrong. Not all parts of our body, but only reproductive cells such as oocytes, contribute inherited material to the next generation.

21. See Mayr (1982), 783.

22. This statement is usually attributed to de Vries, and I follow this tradition. It is the closing statement in de Vries (1905), 825. However, de Vries does not claim to be the originator of this statement, but attributes it to Arthur Harris without further reference. Harris makes this statement in a little-noted article, where he declares it a quote but does not provide the source. See Harris (1904). The statement has resurfaced periodically in the literature, for example in the title of research papers like that by Fontana and Buss (1994).

23. Johannsen himself was very careful not to ascribe any physical reality to genes. See Johannsen (1913), 143–46.

24. The opposite of such discrete inheritance is also called blending inheritance.

25. Darwin knew about discrete inheritance but did not think it was very important. See Mayr (1982), 543.

26. Macromutations may be more frequent in plants than in animals. See Theissen (2006).

27. See Goldschmidt (1940), 391. In the eyes of mutationists, such mutations were much more important to evolution than selection. See Mayr (1982), 540–50.

28. The story of the peppered moth is one of the oldest and most-cited instances of "evolution in action" that have been observed within a human life span. See Kettlewell (1973), as well as Cook et al. (2012). Haldane showed how even such rapid genetic change does not require very strong selection. See Haldane (1924).

29. This assessment comes from human diseases, an especially well studied class of traits. About 1 percent of humans are affected by Mendelian diseases that are caused by mutations in single genes, whereas a much larger percentage of the total population is affected by diseases associated with mutations of weak effects in multiple genes, such as hypertension or diabetes. See Benfey and Protopapas (2005).

30. There are exceptions that prove the rule, such as the phenomenon of polyploidization, where the entire genetic material of an organism becomes duplicated, which can result in major phenotypic change. Many crop plants are polyploids.
31. See Huxley (1942).
32. This quote is itself a simplification from the original in Einstein (1934).
33. See Mayr (1982), 400.
34. See Morgan (1932), 177. Curiously, Morgan was an embryologist before he became a geneticist. For a broader discussion see Gilbert (2003).
35. Population genetics and quantitative genetics have gradually become more sophisticated, and allowed that genes contribute in complex nonlinear ways to a phenotype. They also study multivariate phenotypes, phenotypes that cannot be written as single scalar quantities but are represented as vectors. But even these representations cannot encapsulate the true complexity of phenotypes such as the fold of a protein, which is best represented through the atomic coordinates and molecular motions of its amino acids. The formation of this phenotype is completely determined by its genotype, the amino acid chain, yet it is so complex that we still cannot compute it from information in the genotype.
36. The term *enzyme* itself had been coined already in 1877 by the German physiologist Wilhelm Kühne.
37. See Stryer (1995).
38. See Desmond and Moore (1994). The fifth edition is the first to use the phrase "survival of the fittest," which had been coined by the philosopher Herbert Spencer.
39. See Mayr (1982), chapter 19.
40. See Avery, MacLeod, and McCarty (1944).
41. See Watson and Crick (1953).
42. For which Max Perutz and John Kendrew would share the 1962 Nobel Prize in Chemistry.
43. See Benfey and Protopapas (2005). The very earliest techniques to study genetic variation did not yet read DNA directly, but used alternative measures to measure genetic variation, such as the mobility of different variants of a protein in an electric field. See, for example, Lewontin and Hubby (1966).
44. See Kreitman (1983) and Oqueta et al. (2010).
45. See Eng, Luczak, and Wall (2007). Such individuals metabolize alcohol more efficiently into acetaldehyde, which causes the undesired side effects.
46. The notion of a "paradigm shift" leading to incompatible worldviews was immortalized by the historian Thomas Kuhn. See Kuhn (1962).
47. See Nikaido et al. (2011).
48. One of several differences between the English alphabet and the molecular alphabet of DNA is that they contain different numbers of letters and each letter thus carries different amounts of information.

49. The first draft sequence from both the publicly and the privately funded projects was based on DNA not from just one individual but from multiple individuals. In the privately funded project, some of that DNA came from the project leader himself. See Venter (2003).
50. The diseases he refers to are so-called complex diseases like diabetes, caused by mutations in multiple genes (and influenced by environmental factors such as diet), as opposed to Mendelian diseases, which are caused by mutations in single genes.
51. Other molecular interactions, such as those between proteins and DNA that help regulate genes, are also important in signaling, as chapter 5 will point out.
52. Such mathematical descriptions of biochemical systems existed for many decades, since biologists first described enzymes and the rates at which they can catalyze chemical reactions. See Fell (1997). However, in the last decade of the twentieth century, molecular biology embraced such descriptions as essential to understanding biochemical systems in a newly fashionable branch of biology called systems biology.
53. See Sedaghat, Sherman, and Quon (2002) for a mathematical model of insulin signaling, and Draznin (2006) for some hypotheses of the mechanisms behind insulin resistance. Sanghera and Blackett (2012) discuss some of the genetic complexities of type 2 diabetes.
54. As many scientists after Darwin have forcefully argued. See, for example, Dawkins (1997).
55. To my knowledge, the term *genotype-phenotype map* was coined by the Spanish developmental biologist Pere Alberch, who studied macroscopic phenotypes that were too complex to draw this map in molecular detail. See Alberch (1991). However, the idea behind such maps can be found in the work of many others, such as Sewall Wright, one of the founders of the modern synthesis, or the embryologist Conrad Hal Waddington. See Waddington (1959).
56. See Mayr (1982), 304.

CHAPTER TWO: THE ORIGIN OF INNOVATION

1. See Pasteur (1864).
2. See Horowitz (1956).
3. Ibid.
4. Pasteur was aware that these microbes could enter the growth medium from dust grains in the air. See Pasteur (1864).
5. See Cropper (2001), 259.
6. See Sleep (2010).
7. See Sleep (2010) and Delsemme (1998).
8. See Schopf et al. (2002), but also Brasier et al. (2006).
9. See Mojzsis et al. (1996), but also Lepland et al. (2005).
10. See Oparin (1952) and Haldane (1929).

11. Darwin's letter from February 1, 1871, to his friend J. D. Hooker is available as letter 7471 from the Darwin Correspondence Project (http://www.darwin project.ac.uk/entry-7471).

12. In the interest of historical accuracy, I note that the German chemist Friedrich Wöhler first showed that an organic molecule, urea, could be made from inorganic ingredients.

13. See Miller (1953).

14. See Miller (1998).

15. For the reanalysis of the meteorite content, see Schmitt-Kopplin et al. (2010). The comet's impact is documented by the Meteoritical Society at http://www .lpi.usra.edu/meteor/metbull.php?code=16875. See also Bryson (2003).

16. See Sephton (2001) and Radetsky (1998).

17. See Delsemme (1998).

18. Ibid.

19. See Deamer (1998).

20. See Delsemme (1998).

21. See Watson and Crick (1953).

22. It turns out that even DNA can catalyze some chemical reactions, as Ronald Breaker demonstrated in 1994. However, thus far DNA catalysts exist only in the laboratory.

23. It had been hypothesized that RNA might be a catalyst, partly because it can fold into elaborate spatial structures, but the proof was provided in Guerrier-Takada et al. (1983) and Kruger et al. (1982).

24. Other roles of RNA were known as well, such as that of the transfer RNA that loads the ribosome with amino acids, but none as important as that of a catalyst.

25. The notion of an RNA world comes from Gilbert (1986).

26. See Cech (2000).

27. To be precise, this molecule would actually replicate a template, not itself, so at least two molecules are needed to start the process.

28. See Johnston et al. (2001), as well as Zaher and Unrau (2007) and Cheng and Unrau (2010).

29. See Eigen (1971).

30. See Szostak (2012), as well as Eigen (1971) and Kun, Santos, and Szathmary (2005). This is really just a rule of thumb. The needed accuracy depends also on other factors, such as how much worse the unfaithful copies of an RNA replicase are at replication.

31. See Johnston et al. (2001).

32. See Drake et al. (1998).

33. See Kelman and O'Donnell (1995). The precursors are molecules like deoxy-ATP, whose incorporation into newly synthesized DNA requires energy, which is obtained by cleaving two phosphate residues in the precursor.

34. This calculation is based on a replicase with 189 nucleotides, the same length as the polymerase found by Johnston et al. (2001), as well as on an average molecular weight of 340 grams per mole of a nucleotide building block. It takes into account that one replicase molecule is needed to replicate another molecule, which would decrease the doubling rate of a replicase population. The polymerization rate of one polymerization reaction per second is taken from the so-called class I ligase discussed in Ekland, Szostak, and Bartel (1995), but I note that even if this rate were orders of magnitude slower, there would still be an exponentially growing requirement for nutrients.

35. See Szostak (2012).

36. See Miller (1998).

37. See also Martin et al. (2008) and Braakman and Smith (2013). One of the most prescient early views is provided, once again, by J. B. S. Haldane. See Haldane (1929).

38. See Stryer (1995). Enzymes with especially high rates of acceleration include alkaline phosphatase and urease. Some enzymes, so-called promiscuous enzymes, can catalyze multiple reactions, but one of them is usually catalyzed with the highest efficiency. See Stryer (1995).

39. See Wachtershauser (1992), Wachtershauser (1990), Morowitz et al. (2000), Copley, Smith, and Morowitz (2007), Bada and Lazcano (2002), Ycas (1955), and Martin et al. (2008).

40. See Delsemme (1998).

41. See Corliss et al. (1979).

42. Hot springs and geysers are terrestrial hydrothermal vents.

43. More specifically, they are *chemoautotrophic* organisms that build their bodies using inorganic molecules as energy sources, as opposed to *photoautotrophic* organisms—mostly plants—that use light energy. Organisms like us are *heterotrophic,* feeding on organic molecules that have been created by other organisms.

44. See Martin et al. (2008)

45. See Beatty et al. (2005)

46. "The deep hot biosphere," Wikipedia, http://en.wikipedia.org/wiki/Hydrothermal_vent#The_deep_hot_biosphere.

47. See Kashefi and Lovley (2003).

48. See Holm and Andersson (1998), as well as Martin et al. (2008).

49. See Budin and Szostak (2010), as well as Kelley et al. (2005).

50. See Smil (2000). These kinds of metals have served as valuable catalysts to industrial chemists for a long time. The Haber-Bosch process that sustains a third of the world's population, for example, uses iron to create five hundred million tons of ammonium fertilizer every year. See Holm and Andersson (1998), as well as Hsu-Kim et al. (2008).

51. The citric acid cycle is also called the tricarboxylic acid cycle or the Krebs cycle, after the German-born Nobel Prize–winning biochemist Hans Adolf Krebs.

See Braakman and Smith (2013) for some variants on this theme as a possible origin of metabolism.

52. See Morowitz et al. (2000), as well as Braakman and Smith (2013).

53. See Stryer (1995), as well as Smith and Morowitz (2004).

54. What I have described first is the more primitive *reductive* TCA cycle, which uses energy from reduced inorganic molecules and carbon from CO_2 to synthesize precursors for other molecules. In contrast, the *oxidative* TCA cycle in heterotrophic organisms (like us) extracts energy from organic molecules to produce both energy—ultimately, ATP—and building blocks for biosyntheses, as well as the CO_2 waste product that we exhale.

55. See Hugler et al. (2007), as well as Smith and Morowitz (2004).

56. See Zhang and Martin (2006) and Cody et al. (2000).

57. Theoretical treatments of autocatalytic networks include those by Eigen and Schuster (1979) and Kauffman (1986). It is easy to see how the state of a metabolism can be inherited from parent to offspring, but such inheritance is unlikely to be very faithful, for example because it is subject to stochastic fluctuations in the concentrations of metabolites and catalysts among offspring from the same parent. Nucleic acids clearly provide a superior means of faithful inheritance.

58. See Williams et al. (2011), Huang and Ferris (2006), Ferris et al. (1996), and Holm (1992).

59. See Budin and Szostak (2010).

60. See Deamer (1998).

61. See Budin, Bruckner, and Szostak (2009).

62. Curiously, Pasteur rang the death knell for spontaneous creation, but he still believed that a vital force was necessary for fermentation, which Buchner later showed to require only inanimate enzymes.

63. The numbers I cite here are taken from well-studied cells, such as those of the bacterium *E. coli*. See Neidhardt (1996) and Feist et al. (2007). Although the chemical composition of biomass and thus its building blocks vary among organisms, some important principles hold broadly, such as that proteins, RNA, and DNA typically constitute the majority of biomass.

64. Our human metabolism is even more complex. It has more than two thousand reactions and more than two thousand small molecules. Current knowledge about the *E. coli* network is summarized in Feist et al. (2007), and about the human network by Duarte et al. (2007). Both bodies of knowledge will undoubtedly grow in the future.

65. More precisely, the microbes in our gut synthesize biotin.

66. See Wolfenden and Yuan (2008). I note that sucrase, like other enzymes, does not float through a cell's interior, but is anchored to the membrane of intestinal cells.

67. Some reactions are catalyzed by more than one enzyme, and some enzymes catalyze more than one reaction.

68. To be precise, sucrase is a protein that consists of two identical polypeptides. See Sim et al. (2010).
69. This holds for metabolic enzymes. There are other enzymes, most notably protein kinases, that add phosphates to other proteins, which are large molecules.
70. See Tanenbaum (1988), 254.
71. To be precise, there are several related molecules that can also serve to store energy, such as GTP and deoxy CTP, but they are very similar in chemical structure to ATP, and they use the same kind of chemical bond for energy storage.
72. There are many different kinds of lipids, and membranes vary in their lipid content among organisms, but the principle that membrane molecules are amphiphilic remains unchanged.
73. Some organisms show minor variations from the genetic code. See Knight, Freeland, and Landweber (2001), but these variants most likely originated after the most recent common ancestor of all extant life.
74. One of the alternatives to ATP is its close relative GTP, and one alternative to DNA is PNA (peptide nucleic acid). See Nelson, Levy, and Miller (2000). Chapter 3 of Wagner (2005b) reviews some relevant literature on the genetic code. One of the alternatives may be superior, but that's beside the point. Even if natural selection has caused the demise of the others, the current standards tell us that we descend from a single ancestor.

CHAPTER THREE: THE UNIVERSAL LIBRARY

1. This analogy is inspired by a famous short story of the Argentine author Jorge Luis Borges entitled "The Library of Babel" (Spanish original: "La biblioteca de Babel"), published in English translation in Borges (1962). The idea behind this short story, however, predates Borges. It has been used by many other authors, including Umberto Eco and Daniel Dennett.
2. The BioCyc database can be found at http://biocyc.org/ and is described in Caspi et al. (2012). For the KEGG database see Ogata et al. (1999). Yet another relevant database is described in Chang et al. (2009).
3. See McCarthy, Claude, and Copley (1997), Ederer et al. (1997), Nohynek et al. (1996), Copley (2000), and Copley et al. (2012)
4. See Copley (2000).
5. See Rehmann and Daugulis (2008).
6. See van der Meer et al. (1998) and van der Meer (1995).
7. See Dantas et al. (2008).
8. See Takiguchi et al. (1989).
9. See Mommsen and Walsh (1989) and Wright, Felskie, and Anderson (1995).
10. Plants themselves respire some of the oxygen they produce to build biomass.
11. Salt-loving bacteria also have other adaptations. See Postgate (1994).
12. See Steppuhn et al. (2004).

13. See Bennick (2002).
14. See McMahon, White, and Sayre (1995).
15. The reason is that genomes as similar as these, especially in higher organisms, usually encode metabolisms that are also very similar and do not contain very different sets of enzymes.
16. See Shrestha et al. (2011). They have a mutation that inactivates the enzyme.
17. See Redfield (1993) and Dubnau (1999).
18. The genetic material of some viruses is RNA and not DNA, but their life cycle usually involves a DNA intermediate to which similar principles apply.
19. Excessive DNA can also cause problems when replicated genes or chromosomes need to separate during reproduction.
20. See Bushman (2002), Loreto, Carareto, and Capy (2008), and Bergthorsson et al. (2003).
21. The analogy to human races must be taken with a grain of salt. Bacteria do not reproduce sexually like many animals and plants. The notion of a species is not clearly defined for them, and the same holds for even less precise categories such as race.
22. See Lawrence and Ochman (1998), Blattner et al. (1997), Ochman and Jones (2000), and Pal, Papp, and Lercher (2005).
23. See Lawrence and Ochman (1998).
24. Some relevant articles are Smillie et al. (2011), as well as Ochman, Lerat, and Daubin (2005) and Ma and Zeng (2004).
25. The actual percentage varies, being greater in bacteria and smaller in most multicellular organisms.
26. See Blattner et al. (1997) and Feist et al. (2007). These may also include enzymes that catalyze chemical reactions outside metabolism, such as enzymes that are involved in transmitting information between cells.
27. This simple description hides many technical complexities. For example, even similar genes can sometimes encode enzymes that catalyze different reactions, and vice versa. Also, some enzymes can catalyze more than one reaction, some reactions are catalyzed by multiple enzymes, and some enzymes are the products of not just one but multiple genes. In practice, annotating the metabolic reactions in a genome thus involves more than just computerized comparison of genes. See Feist et al. (2009).
28. This notion of distance is different from the pairwise Hamming distance of two bit strings, which designates the number or fraction of bits at which two strings differ. Specifically, it does not take into account all the reactions that are absent in both metabolisms. Most known metabolisms comprise only a small fraction of the total number of reactions in the known reaction universe. See Ogata et al. (1999). Even if two metabolisms differed in all their reactions, however, there would still exist many reactions that are absent in both metabolisms. For this reason, and because I focus on the proportion of reactions unique to one

network, the fraction of shared reactions, D, is a more appropriate distance measure than the Hamming distance.

29. This becomes less surprising if one is aware that the DNA of two such strains may differ in more than one million nucleotides.

30. I performed this analysis for one bacterial species from each genus, to avoid overrepresenting highly similar species. See Wagner (2009a).

31. While many colors are caused by pigments, in others a finely textured surface brings forth colors through iridescence, such as in the wing coloration of butterflies. In some coloration phenotypes, such as that of the chameleon, both structural colors and pigment-based colors account for the phenotype.

32. Although the procedure I described is feasible, it turns out not to be the most efficient way to compute viability. In practice, an approach called flux balance analysis is more useful. It relies on a computational technique called linear programming. For an overview see Price, Reed, and Palsson (2004). Computations like this can determine more than just viability. They also tell us how fast a metabolism works—how speedily it manufactures biomass molecules. In other words, they can tell us whether an organism could go forth and multiply, or whether it would barely hang on to life.

33. Furthermore, flux balance analysis can also correctly predict growth and nutrient uptake rates under different growth conditions and environments. See Feist et al. (2007), Segre, Vitkup, and Church (2002), Edwards, Ibarra, and Palsson (2001), and Neidhardt (1996). Where predictions and experiment disagree, two principal causes are at work. The first is missing information about a metabolism. The second involves regulatory constraints, where genes for a particular enzyme-catalyzed reaction exist in a genome, but the enzyme is not produced, because the gene is not regulated appropriately. These kinds of constraints are quickly overcome, even in laboratory evolution experiments, and thus do not present a serious obstacle for metabolic innovation. See Fong and Palsson (2004), Fong et al. (2006), Forster et al. (2003), Segre et al. (2002), and Edwards and Palsson (2000).

34. A notable exception would be endosymbionts, organisms that live inside other organisms and benefit from the constant environment their hosts provide. An example of a long-standing endosymbiosis that has endured for many millions of years is found in the bacterial genus *Buchnera*, an endosymbiont of aphids. See Moran, McCutcheon, and Nakabachi (2008), as well as chapter 6.

35. See Feist et al. (2007).

36. A (hypothetical) metabolism viable on all possible fuels certainly has a phenotype, but it can no longer experience a fuel innovation, such that the number of possible innovations must be strictly smaller than that of phenotypes.

37. A classical work exploring spaces of many dimensions is Abbott (2002). A more contemporary exploration can be found in Stewart (2001).

38. In mathematical language, our three-dimensional space and the metabolic library—a space of metabolic genotypes—are both metric spaces, because a notion of distance exists in both of them. See Searcoid (2007). Mathematicians also study nonmetric spaces, but their properties are more difficult to understand intuitively, precisely because they lack a notion of distance.

39. $(4 \times 10^9 \text{ [yr]}) \times (365 \text{ [d/yr]}) \times (8.64 \times 10^4 \text{ [s/d]}) \times (5 \times 10^{30}) = 6.3 \times 10^{47}$ combinations.

40. By conventional measures of scientific output, such as citations of scientific publications per capita, it is arguably the world leader. See Cole and Phelan (1999). For what it's worth, Switzerland has also produced more Nobel laureates per capita than even the United States. "List of Nobel laureates by country per capita," Wikipedia, http://en.wikipedia.org/wiki/List_of_Nobel_laureates_by_country_per_capita.

41. More precisely, it is the ability to synthesize the carbon backbone of all these molecules from the carbon atoms and from the energy stored in this sugar.

42. Other researchers focused on a different question, namely whether all reactions in a metabolism are essential, and found that they are not, which leads to the same conclusion. See Edwards and Palsson (2000), as well as Fong and Palsson (2004).

43. This work was carried out in collaboration with Areejit Samal and Olivier Martin. See Samal et al. (2010).

44. The work I discuss here is summarized in Rodrigues and Wagner (2009), as well as in Samal et al. (2010) and Rodrigues and Wagner (2011).

45. See Rodrigues and Wagner (2009).

46. I note that in our analyses we started from viable networks of different numbers of reactions, and kept the number of reactions in a network approximately constant during a random walk. Each such walk thus explored a "slice" through the hypercube of the metabolic library.

47. See Rodrigues and Wagner (2009).

CHAPTER FOUR: SHAPELY BEAUTIES

1. Fletcher, Hew, and Davies (2001) present an overview of antifreeze proteins in fish.

2. Some enzymes can catalyze multiple reactions, and are often called "promiscuous" for that reason. See O'Brien and Herschlag (1999). Conversely, some reactions are catalyzed by multiple enzymes.

3. See Zhao et al. (2001). Charcot-Marie-Tooth disease can also be caused by mutations in other genes.

4. Other factors matter as well, such as the electric charge of amino acids, but because most of what I say holds for these factors as well, I will let shape stand for them. Binding of molecules with shape complementarity involves specific interactions and attractive forces between molecules, such as hydrogen bonding. See Branden and Tooze (1999).

5. The technically more precise term for a single amino acid chain is *polypeptide*. A protein can consist of one polypeptide or multiple polypeptides.

6. The words I use here to describe protein folding are anthropomorphic, but the process is purely physical, no less than how iron filings align in a magnetic field, but more complicated than that, because multiple conflicting attractive and repulsive amino acid interactions are at work.

7. More precisely, these elements of a protein's structure are called an α-helix and a pleated β sheet. A pleated β sheet forms from parts of the amino acid chain that are not necessarily contiguous in the chain. These parts are also called β-strands. In the figure, they correspond to the nearly straight ribbons terminated by arrowheads. See also Branden and Tooze (1999).

8. What you see is actually only about half of the entire amino acid string, also called the N-terminal domain. The entire sucrase molecule is a complex of two amino acid strings. See Sim et al. (2010). The large size of a protein may provide it with stability to thermal motion, specificity for its target molecule, high rate of catalysis, and the ability to regulate its activity. Although one can synthesize catalytic peptides, much smaller enzymes consisting of few amino acids, such peptides do not have these properties of more complex enzymes. See Tanaka, Fuller, and Barbas (2005).

9. This jiggling is also the basis for enzyme promiscuity, the phenomenon in which some enzymes can catalyze multiple chemical reactions. Some of the oscillating shapes they form can bind molecules other than their main targets, and help these molecules react. They may not be very good at these side jobs, but good enough to accelerate the rate at which these other reactions proceed. See, for example, O'Brien and Herschlag (1999). In the evolution of enzymes early in the history of life, some enzymes probably were highly promiscuous. They catalyzed multiple reactions, each at a low rate, and became specialized later for one reaction that they could catalyze efficiently. See Kacser and Beeby (1984). Because no one enzyme can catalyze all reactions necessary to sustain modern life, promiscuity does not eliminate the need to understand how enzymes with new catalytic abilities arose in the first place.

10. See Szegezdi et al. (2006).

11. From the closing paragraph of Darwin (1859).

12. See Cheng (1998) for a survey on the evolution of diverse antifreeze proteins. The ancestors of these proteins include enzymes and lectins, which have a variety of roles, including the promotion of cell adhesion.

13. The species in question are sculpins in the genus *Myoxocephalus*. See Cheng (1998).

14. Fletcher et al. (2001). Different antifreeze proteins within the same organism might not have a different origin.

15. The ability to protect against freezing may not have arisen abruptly but gradually, where some amino acid changes increased a protein's ability to

protect against freezing to a small extent, until today's antifreeze proteins had formed.

16. The reactions in question are catalyzed by the enzymes HisA and TrpF. See Wierenga (2001).

17. The *E. coli* enzyme is called L-ribulose-5-phosphate 4-epimerase. After the mutation, it becomes an aldolase. See O'Brien and Herschlag (1999).

18. Several other changes occurred in its hemoglobin, but this one, a proline-to-alanine substitution, is especially important. See Liang et al. (2001), as well as Liu et al. (2001) and Golding and Dean (1998). A number of additional mechanisms facilitate high-altitude adaptation. See Liu et al. (2001), as well as Monge and Leonvelarde (1991).

19. Gene duplications of the genes encoding opsins were also involved. Golding and Dean (1998) provide an overview of these and other adaptations.

20. The phenomenon is also known as the Red Queen effect, a term coined by the American biologist Leigh Van Valen.

21. These proteins do not appear out of nowhere. They are modifications of so-called ABC transporters, a very large and widespread class of proteins that transport all manner of molecules in and out of cells, in organisms ranging from bacteria to humans. See Putman, van Veen, and Konings (2000), as well as Gottesman et al. (1995). The modifications can affect the transporter's amino acid sequence or the amount of protein itself, for example by changing the number of genes encoding a transporter in the genome. See Mrozikiewicz et al. (2007), as well as Stein, Walther, and Wunderlich (1994). For the rapid spreading of drug resistance see Tomasz (1997) and LeHello et al. (2013).

22. One could say that these changes are not dramatic on the level of protein phenotypes, but they are dramatic for the (physiological) phenotypes of whole organisms. Thus whether a change constitutes an innovation depends on the level of organization at which one chooses to study the phenotype.

23. Strictly speaking, this genotype is the DNA sequence encoding the protein, but the two are equivalent for my purpose, because a single DNA string uniquely specifies an amino acid string.

24. As of this writing, experiments have determined the folds of more than seventy thousand proteins, and computational methods that infer the fold of one amino acid string from an experimentally determined fold of another, similar string, can infer the shapes of millions more. A central public repository for information about protein fold and function is the Protein Data Bank (http://www.pdb.org).

25. See Maynard-Smith (1970).

26. Many proteins are complexes of multiple polypeptides. Such complexes can be many times larger than any one polypeptide.

27. More precisely, this space is called a generalized hypercube. See Reidys, Stadler, and Schuster (1997). One can walk away from each vertex of this hypercube in

as many directions as the hypercube has neighbors. For a protein of one hundred amino acids, for example, which has nineteen hundred neighbors, nineteen hundred such directions exist.

28. A variety of distance measures exists in sequence space. Several of them take into account that some amino acids are more similar in their chemical properties than others. How far a genotype network reaches through sequence space may vary somewhat with the distance measure used.

29. See Eco (1977) and Putnam (1975) for a discussion of basic semiotic concepts and ambiguities about the meaning of "meaning."

30. Estimates of the fraction of foldable proteins vary broadly between 0.01 and 10 percent. See Keefe and Szostak (2001), as well as Finkelstein (1994) and Davidson and Sauer (1994). For the purpose of this section, I equated meaningful proteins with foldable proteins, because function requires folding in most proteins, with the caveat that some unstructured proteins may also perform useful functions.

31. I use the notion of "work" here in the physical sense.

32. See Keefe and Szostak (2001).

33. Bacteriophages that can go dormant like this are called lysogenic. Their DNA becomes part of the host genome until the host experiences severe stress, at which time the viral DNA starts to express its genes and viral particles are made. See Ptashne (1992).

34. See Reidhaar-Olson and Sauer (1990) and Taylor et al. (2001). Note that even though the total number of sequences adopting a given function may be large, the fraction of sequence space they occupy may be vanishingly small.

35. Although these solutions may differ in their amino acid sequence, they may have other commonalities, for example a particular spatial arrangement of specific amino acids that allows catalysis of a reaction.

36. Our genomes encode more than one globin. The hemoglobin protein itself is made up of four globin polypeptides, two so-called alpha chains and two beta chains, each of which is encoded by different genes. Other globin genes in our genome include one that is mostly expressed during development in the womb, and yet another that is important for binding oxygen in muscles.

37. This is an estimated rate of mutation per human generation, not per round of DNA replication, which would be even lower. See, for example, Nachmann and Crowell (2000).

38. Hemoglobin-related diseases are well studied and known as hemoglobinopathies. Sickle-cell anemia is one of them. Not all of these diseases are caused by alterations of single DNA letters. They can also be caused by deletions of DNA and other genetic changes. Some mutations in the DNA letter sequence of a gene may not affect the amino acid sequence of the encoded gene at all, because the genetic code is redundant, such that some nucleotide combinations encode the same amino acid, a fact that I briefly mentioned in chapter 1.

39. They are taken from the beta chain of hemoglobin.
40. Assuming a human generation time of twenty-five years.
41. The estimates of times to most recent common ancestry I provide are approximate, as these times can only be estimated with substantial error. See, for example, Hedges and Kumar (2004), as well as Hedges and Kumar (2003).
42. Even globins from organisms as different as plants and animals probably were not independent inventions but derive from a common ancestor. See Hardison (1996).
43. A subtle philosophical question is what constitutes different solutions to the same problem. A chemist might argue that two proteins differing in their amino acid sequence but cleaving a small molecule with the same reaction mechanism are similar solutions, whereas two proteins that use a different reaction mechanism are different solutions. From an evolutionary perspective, however, it is sensible to view all genotypes that serve the same function as different solutions to the same problem, because each of these phenotypes can, in principle, be discovered independently from other such genotypes.
44. See Kapp et al. (1995) and Goodman et al. (1988). To this day, proteins may keep diverging further and further from their common ancestor. See Povolotskaya and Kondrashov (2010).
45. I emphasize the role of globins in nitrogen fixation here, but globins can also help distribute oxygen in plants. See Hardison (1996).
46. See Rizzi et al. (1994).
47. See Wierenga (2001). Proteins with this fold can actually have different functions, but even proteins with this fold and the *same* function can be highly divergent. TIM barrels may have originated multiple times independently in the history of life.
48. The argument is analogous to the one from chapter 3 about the exploration of the metabolic library: A few nonillion organisms exploring a new protein every second since life's origins would yield only a vanishingly small fraction of all proteins. It would not even make a difference if this estimate were off by several orders of magnitude.
49. Other factors, such as gene duplication and phenotypic plasticity, can also facilitate innovation in proteins. For an overview of such factors see Wagner (2011).
50. Most of the protein pairs he analyzed were far apart in genotype space, but not so far that they would not have originated from a common ancestral protein, as opposed to having originated independently. See Ferrada and Wagner (2010).
51. RNA can also carry out other functions inside cells, such as to regulate genes through a process called RNA interference. Here, RNA can have an advantage over proteins, because the principle of base complementarity allows it to bind other nucleic acids with high specificity, such as parts of a messenger RNA transcribed from a gene. Among other functions of RNAs, their role in protein

transport is especially noteworthy. It involves the signal recognition particle, an RNA-protein complex that helps proteins enter a part of the cell called the endoplasmatic reticulum.

52. We know the folds of some very well studied molecules, such as the ribosomal RNA that catalyzes the key reaction of protein synthesis in the ribosome, but such information is lacking for many other RNAs.

53. Together with Manfred Eigen, Schuster showed theoretically how heterogeneous populations of RNA molecules that can catalyze each other's production can form self-sustaining systems they called hypercycles. See Eigen and Schuster (1979).

54. These are described in multiple publications beginning with Hofacker et al. (1994).

55. The base pairs that can form in the secondary structure are A-U, C-G, and G-U. (RNA contains the base uracil, abbreviated by the letter U, instead of the base thymine of DNA.) One difference between the helices of proteins and those of RNA is that the helices of protein structures are formed by a contiguous amino acid strand, whereas the helices of RNA are formed by different, generally noncontiguous parts of the same molecule. Many RNA molecules also require interactions with metal ions to form stable tertiary structures.

56. As reported by Schuster et al. (1994), the number S of RNA secondary structures scales exponentially with sequence length L, as $S \propto (L^{-15})(1.85)^L$.

57. An important early paper from Schuster's research group is Schuster et al. (1994), and a broader range of later work is summarized in Schuster (2006). Although based only on secondary structure, this work provides the most comprehensive characterization of a genotype space to date. On a historical note, the first work that provided potential evidence for the existence of genotype networks came before Schuster's, and used simple models of protein folding. See Lipman and Wilbur (1991), as well as Lau and Dill (1989). Like RNA secondary structure models, these models tell us little about the evolution of protein function, and more about the evolution of structure. Schuster's group coined the term "neutral networks" for genotype networks. Although widely used, the term "neutrality" has a specific meaning to most students of molecular evolution. Namely, it implies changes that do not affect fitness in any way. The kinds of changes that distinguish neighboring genotypes on a genotype network are not necessarily of this nature, as I discuss in Wagner (2011). Thus it is best to use this term sparingly, and here I avoid it altogether for this reason.

58. All these observations refer to typical shapes. There may be shapes formed by only a single RNA sequence, but these shapes would be very hard to find in a blind evolutionary search. The vast majority of RNA sequence space is filled with structures that are formed by many sequences. Moreover, the shapes of multiple biologically important RNA molecules are also formed by many sequences, as we were able to show in Jörg, Martin, and Wagner (2008).

59. Schultes and Bartel (2000).

60. To be precise, in these walks they changed two residues at a time to preserve RNA secondary structures. One can think of such changes as a combination of a nucleotide change that disrupts secondary structure, followed by another one that restores it. Such pairs of mutations where one compensates for another are observed quite frequently in nature, and thus occur—although perhaps not simultaneously—in naturally evolving RNA molecules. See Kern and Kondrashov (2004). It is also relevant that these researchers had some inkling that their effort might be successful: They had managed to design a sequence that was intermediate between the starting enzymes and that had both activities.

61. Their work also showed that a sequence that is intermediate between the starting fuser and splitter sequences can catalyze both reactions. See Schultes and Bartel (2000). Such phenotypic plasticity or promiscuity of RNA molecules further helps innovation, because it can make transitions between two genotype networks even easier. See Wagner (2011), chapter 13.

62. The RNA polymerase that copies information in genes is a DNA-dependent RNA polymerase. (It uses a DNA template.) The RNA polymerase that replicates RNA is an RNA-dependent RNA polymerase.

63. It is the so-called group I intron of a transfer RNA gene for isoleucine in the bacterium *Azoarcus*. See Tanner and Cech (1996), as well as Reinhold-Hurek and Shub (1992).

64. The best-known form of the second process involves RNA and not DNA. It is called splicing and occurs when eukaryotes delete parts of a messenger RNA and splice the rest together to create a contiguous stretch of RNA that encodes a single polypeptide.

65. See Hayden, Ferrada, and Wagner (2011). I simplified the description of several aspects of this experiment for brevity. Because of how the experiment was designed, the numbers of molecules fluctuated during different stages of the experiment between one hundred million molecules (after selection) and more than a trillion (10^{12}) molecules. Also, during each generation, each molecule may replicate not just once but multiple times. An important aspect of the experiment's first part was that the activity of the enzyme did not improve, nor did it deteriorate. The population only spread through genotype space without changing its phenotype. The population showed what geneticists call cryptic variation, variation that one cannot normally detect on the level of phenotypes, but that can become visible in a new environment, which involved in our experiment a change in the chemical target molecule of the RNA enzyme. In other words, if our experiment had generated phenotypic variation and this variation had fed the evolutionary process, we would not have been surprised—that is the standard Darwinian view—but the fact that cryptic variation can help evolutionary adaptation is more surprising, and best explained through

the genotype network framework. The new substrate in the experiment's second part could also be transformed by the original enzyme, but at a much slower rate. In other words, the experiment focused on how fast the average reaction rate of ribozymes in the two populations increases during laboratory evolution. In the first population, the rate increased up to eight times faster than in the second population.

66. See Keats (1994).
67. See also Dawkins (1998), where the author points out that there can be wonder and awe even in an unwoven rainbow.

CHAPTER FIVE: COMMAND AND CONTROL

1. See Swallow (2003), as well as Bersaglieri et al. (2004) and Tishkoff et al. (2007).
2. Any genetics textbook, such as Lewin (1997), describes some of their ingenious experimental tricks.
3. There are several kinds of polymerases. The one I am discussing here is a DNA-dependent RNA polymerase.
4. Strictly speaking, the RNA sequence is complementary to one strand of the gene and identical to the other strand, because DNA is a double-stranded molecule.
5. β-galactosidases are enzymes that cleave the sugar β-galactose from larger sugars. Because lactose is one such sugar, lactase is a kind of β-galactosidase. For historical reasons, the gene encoding E. coli's β-galactosidase is called the lacZ gene. See Lewin (1997), chapter 12.
6. More precisely, the complete word is TGTGTGGGAATTGTGAGC-GATAACAATTTCACACA, and the regulator does not make specific contact with all the letters in this sequence. See also Lewin (1997), chapter 12. The word is very similar to a palindrome, a DNA word that when read in one direction on one strand gives the same letter sequence as read in the reverse direction on the opposite strand.
7. For example, there may be up to twenty thousand elementary protein shapes or so-called domains. See Levitt (2009).
8. The details of regulation are more complicated than I describe. For example, the regulator (called the lac repressor) is actually a complex of four polypeptides. And it regulates the expression of not just one but three adjacent genes, the so-called lac operon. But all these details leave the principles of regulation unchanged. See Lewin (1997), chapter 12.
9. See Russell (2002), chapter 16. Yet another cost factor is that synthesizing a useless protein ties up ribosomes—the large complexes of molecules that translate proteins from RNA—which are thus not available to synthesize other, necessary proteins.
10. See Dekel and Alon (2005).

11. The interaction between activator and polymerase need not be direct. For example, an activator bound to DNA may change the conformation of DNA to open its double helix and thus make it easier for polymerase to start transcription. Nonetheless, the principle of complementarity is also important in transcriptional activation.

12. With the exception of a few types of cells, such as red blood cells, which have shed their genome.

13. See Poole et al. (2001), as well as Piatigorsky (1998) and Morano (1999).

14. I am referring to type II collagen, encoded by the human COL2A1 gene. Other tissues, such as skin or hair, contain other types of collagen, made by different genes. Regarding motor proteins, I am referring to myosins, which are encoded by a large family of closely related genes in the human genome. Different tissues express different members of this family. Not all of them serve to contract muscles. Some, for example, transport molecules inside a cell.

15. In the evolution of some new cell types, multiple kinds of innovations may be involved, for example because some new cell types require both new regulation of existing proteins and new proteins.

16. See Gilbert (2010), 42–43, for the regulation of the chick δ1 crystallin gene by Pax6, and for the regulation of Pax6 itself. Pax6 stands for "paired box 6," which refers to a 120-amino-acid-long element of the protein's structure that is similar among Pax6 proteins of different organisms.

17. I simplify again for brevity. For example, the polymerase itself is not one protein, but a complex of more than a dozen proteins, some of which can interact with transcriptional regulators, whereas others cannot. Each transcriptional regulator may also comprise more than one protein, and there may be more than one binding site—and sometimes many—for any one regulator near the same gene. Some DNA-bound regulators act alone, whereas others need to physically interact with other proteins to regulate transcription.

18. For Pax6 and its role in eye disease and development see Hingorani, Hanson, and van Heyningen (2012), as well as Tzoulaki, White, and Hanson (2005) and Ashery-Padan et al. (2000). For its relationship to eyeless and fruit fly eye development see Gehring and Ikeo (1999).

19. A wiring diagram is an abstraction that allows our eyes to grasp a circuit's genotype at a glance. It is easily augmented through equations that describe the interactions of circuit genes encoded in DNA in great detail, but such equations are less easy on the eyes.

20. Some gene expression patterns may not reach an equilibrium but may vary cyclically, which can be important to maintain rhythmic behavior, such as the circadian clocks that regulate day-night activity cycles.

21. The prize was awarded to Christiane Nüsslein-Volhard, Edward B. Lewis, and Eric Wieschaus. See Lewis (1978) and Nüsslein-Volhard and Wieschaus (1980). Development does not sculpt bodies in the same way for all insects. For

example, in fruit flies all segments are laid down early in the embryo, whereas in grasshoppers, posterior segments form in a posterior proliferation zone during development. Thus any one species can serve as a model of development only for a restricted group of other species.

22. To be precise, in the early fly embryo not cells but nuclei divide, which makes the spreading of signals among them easier. And here are some further simplifications: Diffusion and transport of some RNA molecules is facilitated by cytoskeletal elements. Some RNA molecules, such as that of the regulator *hunchback,* do not show a gradient like bicoid but are uniformly distributed throughout the embryo. Hunchback's translation is inhibited by another regulator, *nanos,* concentrated at the posterior pole, which contributes to a hunchback gradient. Moreover, while all these molecules are regulators, not all of them are transcriptional regulators. Nanos, for example, affects the translation of other RNAs.

23. See Carroll, Grenier, and Weatherbee (2001).

24. At this stage, nuclei have become separated by cell membranes, regulators can no longer diffuse freely through the embryo, and cells need to communicate in ways that can overcome membranes.

25. The actual sequence of events is again more complicated. Upon hormone binding, so-called heat shock proteins bound to the receptor dissociate from it, the receptor gets transported into the cell nucleus, and it dimerizes, that is, it forms a complex of two proteins.

26. This might happen if the regulator gene suffers a mutation. See Reinitz, Mjolsness, and Sharp (1995), as well as Jäger et al. (2004) and Mjolsness, Sharp, and Reinitz (1991).

27. I here focus mostly on one aspect of limb formation, that is, how limbs get structured along their proximodistal axis—the one that extends from the part of a structure closer to the body to the part farther away from the body. There are several other well-studied questions, such as how limbs "know" where to sprout along the head-tail axis, and how they are patterned along the axis running from back to front, the dorsoventral axis. See Carroll et al. (2001).

28. See Lewis (1978). See also Gilbert (1997), chapter 14.

29. See Cohn and Tickle (1999). An expanding region of thoracic identity is only part of the explanation for why snakes have more (thoracic) vertebrae. The other comes from a segmentation clock—driven by yet another regulation circuit that determines segment and vertebrae number. See Gomez and Pourquie (2009).

30. See Davis, Dahn, and Shubin (2007), as well as Sordino, Vanderhoeven, and Duboule (1995).

31. See Zakany and Duboule (2007). For Hox genes and human limb malformations see Goodman (2002). Human birth defects involving Hox genes are rarely as clear-cut as these examples, because of partial redundancies involving several Hox genes.

32. See Stevens, Stubbins, and Hardman (2008), as well as Stevens, Hardman, and Stubbins (2008) and Stevens (2005), for the adaptive role of eyespots. The gene *distalless* derives its name from the fact that mutations in it eliminate the distal-most part of a fly's legs, that is, the part farthest away from the body. See Panganiban and Rubenstein (2002), as well as Dong, Dicks, and Panganiban (2002) and Carroll et al. (2001).

33. See Brakefield et al. (1996) and Keys et al. (1999).

34. See Kenrick (2001), as well as Beerling, Osborne, and Chaloner (2001) and Gottschlich and Smith (1982). Not all plants have dissected leaves, because leaf dissection has a price: A greater surface area also means greater water loss, which is a disadvantage in dry climates.

35. See Bharathan et al. (2002).

36. See Hay and Tsiantis (2006) and Bharathan et al. (2002). Strictly speaking, KNOX, which stands for class I KNOTTED1-like homeobox proteins, is not just one regulator but a family of similar regulator proteins that act in different plants. I also note that dissected leaves may have originated through additional changes in one or more of the other genes in this circuit, but none of that takes away from the importance of regulatory change for their origin.

37. See Babu et al. (2004).

38. Note that pairs (A,B) and (B,A) need to be distinguished, because gene A can regulate B differently from how gene B regulates gene A. In addition, this procedure allows for the possibility that a gene can regulate itself, which is often the case.

39. The number of Hox genes varies even among chordates and vertebrates. For example, the basal chordate *Amphioxus* has only ten Hox genes, whereas some fish species have many more than forty Hox genes, which is a result of a past duplications of their entire genome. See Amores et al. (1998) and Garcia-Fernandez and Holland (1994).

40. Much of the work that I summarize in this chapter can be found in Ciliberti, Martin, and Wagner (2007a), as well as in Ciliberti, Martin, and Wagner (2007b) and Martin and Wagner (2008). The message of this chapter holds even when the strengths of regulatory interactions can vary continuously.

41. In metabolic genotype space, each hypercube vertex corresponds to a binary string that represents a metabolic genotype. A complication for regulatory circuits is that even in the simplest case where regulation can only be activating, repressing, or absent, the relevant strings are no longer binary but trinary—each digit can assume three values. The resulting hypercubes are even less intuitive geometrically, and are sometimes called generalized hypercubes. See, for example, Reidys (1997).

42. One might argue that there is not one circuit library but many, each one for circuits with different numbers of genes and thus existing in a different dimension. However, one can always view lower-dimensional spaces as being

embedded in higher-dimensional spaces. This is why I often use the notion of a library in the singular here.

43. It may seem far-fetched to speak of meaning outside the context of human language. But as I mentioned in chapter 4, to do so has a long tradition in the field of semiotics. For a concise introduction to semiotics see Eco (1977). For an exploration of different kinds of meaning in living beings see chapter 2 of Wagner (2009b).

44. It is important to be aware that the evolution of many innovations such as dissected leaves occurred gradually and not in one giant leap. If an increase in leaf dissection is advantageous, then a "weakly" dissected leaf is better to have than a simple leaf, and a strongly dissected leaf is even superior to a weakly dissected leaf. In consequence, the expression codes that mediate leaf dissection may have changed gradually, in small steps, until strongly dissected leaves were fully formed. Such gradual changes may even have occurred in the formation of very complex organs such as eyes—having imperfect eyes is usually better than being blind. See Gerhart and Kirschner (1998), chapter 5, and Land and Fernald (1992). None of this takes away from the observation that new expression codes are much easier to find in circuit libraries that are organized as I describe here. I do not emphasize the gradual notion of evolution, and use the black-and-white distinction between old and new expression codes simply because it illustrates the relevant concepts more clearly.

45. For pertinent work see Espinosa-Soto, Padilla-Longoria, and Alvarez-Buylla (2004), as well as Albert and Othmer (2003) and Jäger et al. (2004). These researchers model the spatial organization of an embryo in greater detail than the work I describe here, because they focus on only one circuit. That level of detail would be prohibitive for current computational technology if one needed to explore many circuits, except for especially simple circuits like those in Cotterell and Sharpe (2010).

46. For example, two different types of circuits pattern the dorsoventral axis and the proximodistal axis of our limbs. See chapter 3 of Carroll et al. (2001). I leave out spatial considerations for brevity here, because they do not affect the main principles I discuss.

47. For evidence that regulatory DNA changes more rapidly than transcriptional regulators themselves see Tirosh et al. (2009), as well as Wittkopp, Haerum, and Clark (2008) and Wittkopp, Haerum, and Clark (2004). Not all detrimental changes in a circuit are immediately lethal to the individual. The vast majority of harmful DNA changes have subtle effects on individuals, and their lethality manifests itself only on evolutionary time scales, when a lineage carrying them gets eliminated.

48. As shown in Stone and Wray (2001). The changes that occur in such regulatory DNA are not necessarily only changes in individual nucleotide letters of DNA. Deletions or duplications of short stretches of DNA can also occur. I note that some changes in the DNA binding sites of transcriptional regulators may have

no effect on regulation, because a regulator may regulate the same gene via multiple different, redundant binding sites.

49. This possibility is not so far-fetched, even though there are more circuits than equilibrium gene expression patterns: Most circuits in the library do not reach a stable equilibrium gene expression phenotype, but one that varies cyclically. See, for example, Ciliberti, Martin, and Wagner (2007b).

50. More precisely, I am referring to the Institut des Hautes Études Scientifiques (IHÉS) in Bures-sur-Yvette, near Paris.

51. On a historical note, mathematical biology, some of which aims to solve similar problems, is a research field with a long tradition in biology. See Murray (1989). Even the notion of systems biology is far from new. See Bertalanffy (1968). However, the mainstream of biology, especially cell and molecular biology, has acknowledged the importance of these ideas only since the late 1990s.

52. A well-known statistical description of a gas is the ideal gas law, which links pressure, volume, and temperature of a known quantity of gas molecules.

53. Our explorations of the library were greatly aided by postdoctoral researcher Stefano Ciliberti.

54. The exact number of neighbors with the same phenotype depends on a circuit's size and the actual phenotype. Even two circuits in which these properties are the same may have different numbers of neighbors with the same phenotype. See Ciliberti, Martin, and Wagner (2007a) and Ciliberti, Martin, and Wagner (2007b). Here and below, I always discuss *typical* features of circuits. Exceptions may exist, but their overall impact on innovability is small, given that most circuits in the library follow the rules rather than being exceptions.

55. See Wagner (2011), figure 3.3. Different phenotypes do in fact have different numbers of circuits associated with them, but this number is typically very large regardless of the phenotype, as long as a circuit has a minimum number of genes. The variation in size among different genotype networks has implications for innovability that are too technical to cover here, but see Wagner (2008).

56. Ciliberti, Martin, and Wagner (2007a).

57. See Isalan et al. (2008). Gene expression phenotypes may change in these circuits, but what the experiment shows is that the circuits continue to function and sustain life.

58. See Martchenko et al. (2007).

59. See Tanay, Regev, and Shamir (2005).

60. For brevity, the narrative assumes tacitly that a new phenotype will replace the old phenotype. That assumption is not necessary. The same developmental regulation circuit can produce different gene expression patterns in response to different chemical signals, and in different regions of a developing embryo. Suppose that a given expression pattern has a well-established role in

structuring one body part, but that a new expression pattern, as yet undiscovered, produced in response to a second chemical signal can help structure a new body part. A population that drifts through the genotype network of the first expression pattern will be able to explore many different phenotypes in response to the second chemical signal, wherever this signal occurs in the body. Thus, even where circuits have multiple signals and multiple expression codes, the circuit space organization I describe facilitates innovation.

CHAPTER SIX: THE HIDDEN ARCHITECTURE

1. See Waddington (1942) for the quotation, as well as Waddington (1953) and Waddington (1959).
2. For early reviews, see Tautz (1992). For some later work with useful references, see Wagner (1999) and Wagner (2005a).
3. The expression of such a gene costs energy, for example in the manufacture of RNA and amino acid building blocks.
4. See Goffeau et al. (1996).
5. Biologists had observed the effects of mutations for many years, and they had also been able to introduce random mutations into a genome, but until the late twentieth century they could not engineer mutations in a highly targeted and specific fashion.
6. See Giaever et al. (2002) and Winzeler et al. (1999) for examples of large-scale knockout studies in the brewer's yeast *Saccharomyces cerevisiae*. The most important such defect one finds in such mutants is slower reproduction, because it's the one that nature punishes immediately, but it's not the only possible defect. Others include a reduced efficiency of mating or spore formation, and a lower chance of survival in stressful chemical environments. I discuss the role of different aspects of fitness and of environmental variation in the interpretation of knockout experiments in Wagner (2011) and Wagner (2005b).
7. And not just to gene deletions, but to a variety of manipulations that reduce gene expression. One of them takes advantage of a natural process called RNA interference, which is able to block the transcription of specific genes into RNA. For a pertinent study in the worm *Caenorhabditis elegans* see Kamath et al. (2003).
8. See Lander et al. (2001).
9. Duplications can affect more than one gene, entire chromosomes, or entire genomes. Also, sometimes the product of a duplication is two genes that are not quite identical. But none of these caveats affect the principles I discuss here.
10. An intriguing question is whether redundancy in living beings ultimately exists *because* it provides protection against mutations. See Wagner (1999) for a relevant study.

11. Analyzing how missing chemical reactions affect metabolism is usually more quantitative than I outlined here. One typically either measures the cell division rate in a population of cells or computes the so-called biomass growth flux. In such an analysis, there is no all-or-none distinction between essential and nonessential genes, as some genes slow down "city traffic" more than others when knocked out. In general, the number of reactions that reduce biomass growth flux to zero is small. In the wild, many microbes grow and divide very slowly in their native habitat. Such microbes may have other advantages, such as better survival when starved for nutrients. This means that not only those metabolisms that support rapid cell growth and division are successful in evolution.

12. For a discussion of the distinction between redundancy and this kind of "distributed robustness" see Wagner (2005a).

13. These lysozymes in different organisms do not have the same genotype. To the contrary, they are highly diverse, once again reflecting the principle that similar molecular phenotypes can be created by very different genotypes.

14. See Kun, Santos, and Szathmary. (2005).

15. Experimental studies often focus on one aspect of a metabolic phenotype, such as viability on the sole carbon source glucose, but computational work demonstrates that other aspects of metabolic phenotypes can be robust to a similar extent.

16. I note that everything I said thus far about robustness pertains to the robustness of a genotype. One can also define the robustness of a phenotype, a concept that will not play an important role in this book, but that can be important to study the relationship between robustness and innovability. See, for example, Wagner (2008).

17. The branch of mathematics needed to prove this statement is graph theory, and specifically the theory of generalized hypercube graphs. For a glimpse at some of its arcana see Reidys, Stadler, and Schuster (1997). For a more accessible exposition see Wagner (2011), chapter 6.

18. See Darwin (1872), chapter 6, page 170. In the same chapter he expresses his faith in selection's power to preserve small and useful improvements.

19. See Land and Nilsson (2002). Refraction is the change in the direction of a wave caused by a change in its speed, when the wave passes from one medium to another.

20. The kind of regulatory change that allows this high expression varies from crystallin to crystallin. Many crystallins have undergone gene duplication, but nonduplicated crystallins also exist. They include ε-crystallin, which is the same as lactate dehydrogenase, and τ-crystallin, which is the same as α-enolase. See Piatigorsky and Wistow (1989), as well as Tomarev and Piatigorsky (1996) and Piatigorsky (1998). In such nonduplicated crystallins changes in regulatory DNA regions allow enhanced gene expression in the lens. See Jornvall et al.

(1993) for crystallins related to alcoholdehydrogenase. For other examples of co-option see True and Carroll (2002) and Keys et al. (1999).

21. For the extraordinary half-life of crystallins see Lynnerup et al. (2008).

22. See Graw (2009).

23. A useful source on the evolution of vision is Eldredge and Eldredge (2008).

24. See Gould (1993) and Gould and Lewontin (1979).

25. As quoted in Burr and Andrew (1992).

26. For the quote and a brief history of the technology see Stewart (2012), chapter 11. See http://pittsburgh.cbslocal.com/station/newsradio-1020-kdka/ for the KDKA radio station.

27. The numbers cited in this section refer to an analysis of the secondary structure of the molecule, which is not only computationally predictable but essential for ribozyme function. The number of secondary structures in the molecule's neighborhood can be smaller than 129 for several reasons, the most important being that several neighbors can have the same shape.

28. How to compute the huge number of RNA molecules with this phenotype is described in Jörg et al. (2008).

29. The general principle is this: If all individuals in a population are confined to exploring new phenotypes from one place in genotype space, their chances of uncovering a superior phenotype are slim, incomparably slimmer than if they can explore many different neighborhoods, as permitted by the existence of genotype networks.

30. More precisely, I am referring here to the typical distance between two arbitrary genotype networks, not to the distance from a specific genotype to one with an arbitrary new phenotype. Furthermore, I note that statements like this always refer to typical cases. More precisely speaking, they hold, as mathematicians say, "with probability one" as a system grows very large in size. See, for example, Reidys et al. (1997) for some of the relevant mathematics. Phenotypes where the properties I discuss do not apply certainly exist, but they are exceptions to the rule defined by typical phenotypes—phenotypes with large genotype networks. And biologically important phenotypes are typical phenotypes. In a research project that speaks to this issue, we computed the sizes of genotype networks for some eighty different biologically important RNA phenotypes. These sizes were not smaller but even greater than those of random phenotypes. See Jörg et al. (2008). In hindsight, this is expected, because phenotypes associated with large genotype networks are easier to find in genotype space than other phenotypes.

31. This volume depends on the number of genes in a circuit, but it is tiny even for circuits with fewer than twenty genes, the example I use here. See Ciliberti, Martin, and Wagner (2007b) for a more detailed analysis of this and other properties of circuit libraries. In general a circle's area and a ball's volume calculate as πr^2 and $(4/3)\pi r^3$, respectively, where r is the radius and $\pi \approx 3.14$. In

higher dimensions, analogous expressions exist, but they are more complex, e.g., the volume of a five-dimensional ball is given by $8\pi^2 r^5/15$. The numbers I discuss in the text can be calculated from formulae like these and a value of $r =$ 0.15, if the squares and cubes at issue have sides of length 1.

32. My earlier note that these are statistical statements applies here as well: This is the typical outcome, with possible exceptions.

33. I note that the value of 0.75 is already close to the maximally possible ratio of "volumes" in two dimensions, because the largest circle that one can inscribe into a square of volume 1 has a radius of ½, and such a circle occupies a fraction 0.785 of the square's area.

34. For relevant observations pertaining to RNA secondary structures, see Schuster et al. (1994). For RNA tertiary structures and their functions, it is not known whether the same holds. However, because secondary structures are prerequisites for tertiary structures, it may be possible to extrapolate these observations to tertiary structures. Proteins, in contrast, are typically more conserved over large genotype distances, thus suggesting that in a small neighborhood of a genotype, it may not be possible to find all possible protein folds. However, this statement applies to a fold as defined by a protein's arrangement of major secondary structure elements, and many more subtle changes can lead to new functions. For example, some proteins with the same arrangement of secondary structure elements have evolved multiple different enzymatic functions. The innovability of protein functions is an area where substantial discoveries are still waiting to be made.

35. Whether a simple device would be more robust to parts failure, simply because it has fewer parts, or whether a more complex device might be more robust, because any one part is less important and because any one configuration has many neighbors that preserve function, may depend on details of a technology and of the design of a device.

36. See Lawrence (1992).

37. See Gierer and Meinhard (1972) for a simple principle that is useful to make patterns such as segmented bodies. This principle may be realized in many systems, but it is often disguised by an enormously complex signaling circuit involved in pattern formation. For the complexity of the insect segmentation network see Akam (1989), as well as Jäger et al. (2004) and von Dassow et al. (2000). I also note that the relevant environment comprises more than just the world outside a fly's body, and also includes the constantly fluctuating concentrations of molecules inside this body. At least some of the complexity of pattern formation exists to buffer development against these kinds of fluctuations. See, for example, Lopes et al. (2008) and Ochoa-Espinosa et al. (2009).

38. See Samal et al. (2010), as well as Gerdes et al. (2003), for an experimental analysis in a different environment.

39. They are also valuable to some species of ants, which milk aphids for the honeydew their body releases, and provide protection in return.

40. It turns out that the endosymbionts are especially heat-sensitive and thus limit the range of habitats that aphids can occupy. See Ohtaka and Ishikawa (1991).
41. A less charitable analogy, one that holds for many endoparasites, is that of a prisoner. *Buchnera* can no longer live on its own. It is completely dependent on what its host provides.
42. For the process of genome reduction in *Buchnera* see Moran, McLaughlin, and Sorek (2009), as well as Moran and Mira (2001), Tamas et al. (2002), and van Ham et al. (2003).
43. See Yus et al. (2009) and Razin, Yogev, and Naot (1998).
44. See Samal et al. (2010).
45. See Thomas et al. (2009).
46. See Rodrigues and Wagner (2011). There are many ways of defining complexity, but a simple definition—the number of reactions in a metabolism or, more generally, the number of parts in a system—suffices for my purpose.
47. See Samal et al. (2010) and Rodrigues and Wagner (2011).
48. Some aspects of an organism's complexity, such as the size and organization of its genome, may also be augmented by deleterious changes whose effect is so weak that natural selection cannot eliminate them, at least in small populations. See, for example, Lynch (2007).

CHAPTER SEVEN: FROM NATURE TO TECHNOLOGY

1. See Sproewitz et al. (2008) and Moeckel et al. (2006).
2. I will here follow Arthur (2009), 27, in using two alternative and complementary definitions of "technology." In the first, technology is a means to fulfill a human purpose. In the second definition, technology is an assemblage of practices and components. In either sense, biotechnology or digital electronics would be technologies.
3. As quoted in Alfred (2009).
4. As quoted in Lohr (2007).
5. Since the beginning of the twentieth century, objections have occasionally been raised to the notion that mutations are blind or "random" with respect to whether they improve or impair a protein's function. But even the most carefully documented objections have not stood the test of time and have eventually been refuted by data. See Cairns, Overbaugh, and Miller (1988), as well as Foster (2000) and Hall (1998). For all we know, nature has no foresight into the effects of genetic changes. Given the many ramifications that such changes have for the short-term and long-term future of an organism, this is not so surprising. Even we humans with our cognitive abilities are notoriously poor at predicting the effects of interventions in complex systems, from proteins, cells, and organisms to ecosystems or financial markets.
6. See Burchfield (1990), 43. In 1897 he settled on somewhere around twenty million years.

7. This statement is sometimes attributed to Max Planck but may be apocryphal.
8. See Rosen (2010).
9. See Merton (1936) and Merton (1968), 477.
10. See Ogburn and Thomas (1922).
11. Merton called a historical predilection for focusing on single inventors the "Matthew effect," from the passage in the Gospel of Matthew that reminds us that "for unto every one that hath shall be given." It is related to Stigler's law, propounded by the statistician Stephen Stigler, who wrote that "no scientific discovery is named after its original discoverer." Stigler then, in a self-referential joke, credited Merton as the "real" discoverer of Stigler's law. See Gieryn (1980), 147–57.
12. Nature's solutions are discussed in Rothschild (2008). Some of them may be superior in certain environments, such as in an aerobic atmosphere, which helps explain why multiple solutions to carbon fixation persist to this day.
13. For many examples of multiple independent innovations in nature see Vermeij (2006). While life has discovered some innovations more than once, it may have discovered others only once, but the genotypes encoding them may have diversified later beyond recognition. In some systems, for example proteins where current genotypes are extremely diverse, it is difficult to distinguish multiple independent origins from a single origin followed by diversification.
14. See Johnson (2010), 153.
15. These and many other examples of combinatorial innovation and reuse of existing objects and technologies can be found in Kelley and Littman (2001) and Arthur (2009).
16. As quoted in Arkin (1998).
17. See Gould and Vrba (1982). Gould and Vrba used the term for changes that conferred a function different from the original function, and for changes that had no utility when they first appeared.
18. See Darwin (1872), chapter 6, page 175. The example Darwin had in mind was the transformation of the fish's flotation bladder into the lungs of terrestrial animals.
19. See Sumida and Brochu (2000).
20. This and several other examples are discussed in True and Carroll (2002), who call the reuse of old parts co-option. I note that Sonic hedgehog is not a transcriptional regulator, but a molecule involved in signaling between cells. For the naming of Sonic hedgehog see Riddle et al. (1993).
21. More precisely, I am referring to an internal combustion air-breathing turbofan engine.
22. See Arthur (2009), 19. The book also discusses the jet engine in some detail.
23. To my knowledge, one of the first who developed this idea for engineering applications was the German Ingo Rechenberg. See Rechenberg (1973). He

pointed out that mutations need to have effects on a system's behavior or performance that must not be too large in order for an evolutionary algorithm to be able to improve it.

24. This algorithm of mutation and selection is actually a stochastic algorithm, where individuals can sometimes survive based on dumb luck. Such stochasticity gives rise to a process called genetic drift that is important in biological evolution. See, for example, Hartl and Clark (2007).

25. There are many flavors of evolutionary algorithms. Two especially prominent ones are genetic programming and genetic algorithms. See Koza (1992) and Mitchell (1998).

26. More commonly known as NP-hard. See Moore and Mertens (2011).

27. An analysis that suggests how bumblebees might solve this problem is given in Lihoreau, Chittka, and Raine (2010).

28. In contrast to traditional (nonevolutionary) algorithms that can solve instances of the traveling salesman problem involving millions of cities within a few percent of optimality, evolutionary algorithms are general-purpose algorithms that can provide good (if not perfect) solutions for less well studied problems, or for problems that are not as clearly posed mathematically.

29. See Dong and Vagners (2004).

30. For an example on engine design see Senecal, Montgomery, and Reitz (2000). One of the principal problems that evolutionary algorithms face is to find a good "genotype" representation of the features to be optimized, and to find mutation or recombination "operators"—the routines that modify the genotype—that work well. But the reason why such algorithms have not revolutionized engine design may lie deeper, in a technology that does not easily admit recombination through standardized linkage. For technologies with such linkage, choosing mutation and recombination operators may also be easier.

31. This does not mean that they do not use recombination, far from it, but they usually recombine abstract (bit-string) representations of candidate solutions to a problem, and not elements of the solution itself, like the amino acids of proteins. More generally, I note that I use the word "recombination" here in a broader sense than prescribed by its standard definition in genetics—the swapping of DNA molecules. This more general use applies to DNA but also to any other notion of genotype, and can even apply to human innovation that is combinatorial. In this sense, even a change in one or a few amino acids of a protein amounts to a recombination of amino acids.

32. More succinctly, new circuits consist of new combinations of interactions between regulators, interactions that may already have existed in other circuits.

33. See the Web site of the Centro Internazionale di Studi di Architecttura Andrea Palladio (CISA) at http://www.cisapalladio.org/veneto/index.php?lingua=i&modo=nomi&ordine=alfa.

34. Their work is described in Hersey and Freedman (1992).
35. An algorithm could produce more than one outcome, more than one floor plan, because individual instructions in the algorithm may have a stochastic component, such as "subdivide a room into two, three, or four rooms with equal probability." Such stochastic algorithms are widespread in computer science.
36. Other rules include that splits are executed such that buildings are generally bilaterally symmetric around a central axis, walls are aligned whenever possible, and no rooms are allowed to be as wide or long as the entire length or width of the building.
37. More than a century after Palladio, the Swedish inventor and industrialist Christopher Polhem invented a "mechanical alphabet" of machines that is described in Strandh (1987). The letters in this alphabet are machine components, such as levers, wedges, screws, and winches. Polhem believed that through combining these machine components, one could build any conceivable mechanical device. He intended the alphabet as a teaching tool, but some models of machines written in this language have been built. While the idea behind this alphabet is important for students of innovable technologies, it is worth pointing out that the links between machine parts are not standardized. In a similar vein, Sanchez and Mahoney point out that the automobile, aircraft, consumer electronics, and other industries build many different products by combining a limited number of "modular" components. See Sanchez and Mahoney (1996). But yet again, the links between these modules are often non-standardized, custom-made, one-of-a-kind. This is an important shortcoming of the "design spaces" of many technologies and a major difference from the genotype spaces of living beings. For the notion of a design space see Stankiewicz (2000).
38. Even more precisely, in mathematics, a function f can be described as a set of ordered pairs (a,b), where a is a member of a set called the domain of the function that defines all admissible function arguments (inputs), and b is a member of the set of output values the function can take. One writes $b = f(a)$. Digital logic circuits compute functions of bit strings whose outputs are again bit strings.
39. Actually it would report even more. The final version of the Köchel catalog—there are six—does end with #626, the *Requiem in D Minor*, but also includes numerous entries with variations, such as the *Church Sonata in C* (#317a) and the *Aria for Soprano in B-flat* (#317b).
40. Some functions, such as the NOT and NOR functions, can be implemented with single transistors. The upper output bit shown in figure 22 corresponds to the lower significance binary digit. It is computed by the so-called exclusive or function (XOR). The lower-output bit is the higher-significance "carry" bit that is the result of an AND function.

41. Integrated circuits also contain some more complex, derived logical building blocks, such as multiplexers, demultiplexers, and registers, but all of these are based on Boolean logic functions. See also Balch (2003).

42. Such devices also come under other names, such as reconfigurable hardware or programmable logic devices. For an overview see Balch (2003). There are several classes of such devices. The research I discuss here was carried out with a specific class in mind: field-programmable gate arrays (FPGAs). Like many other integrated circuits, such arrays contain many more than just AND and OR gates. They also use more complicated units of computation, such as logic blocks, each of which may contain a full adder, several lookup tables that store truth tables in random access memory, multiplexers, and others. But the principle is still the same: They perform complex computations by networking simpler computational units.

    The programming of such devices is different from more conventional software programming. A search program like the one I described earlier to search a music database is typically executed on a hardwired chip, whereas on a programmable chip a program can change the wiring of the chip itself. It amounts to creating a piece of hardware that can execute a given computation task faster than software running on a generic hardwired chip. Limitations of FPGAs include that they are slower and more expensive than application-specific integrated circuits. See also Balch (2003), 250.

43. The field of machine learning is a well-established research field whose current focus is not programmable hardware, but methods (often implemented in software) that enable computers to extract statistical information from complex data.

44. Once again, I am using the word "wires" not in a literal sense—flexible metal threads—but metaphorically. They are integral parts of an integrated circuit.

45. A difference from proteins is that amino acids form a linear string, whereas the gates in a circuit have a two-dimensional arrangement.

46. The number of functions a circuit can compute depends not only on its number of gates but also on the number of input and output bits.

47. It also allowed us to study idealized situations, such as exploring every possible wiring change in a circuit, whereas some such changes may be prohibited in a commercial FPGA architecture. We considered circuits that could harbor OR, AND, XOR, NAND, and NOR gates, because these are the five most commonly used two-input logic gates. Each input could be wired to any of the input gates in the first column of the sixteen-gate array. Each output could be wired to any of the circuit's output bits. Gates were wired internally in a feed-forward fashion where the input of a gate could only come from a gate to the left of it in the array. Such feed-forward connectivity is important to avoid complex dynamics, such as cyclic behavior. More details on this work, including some of the numbers I discuss, can be found in Raman and Wagner (2011).

48. The circuits we studied had four input and four output bits, and there are 1.84×10^{19} possible Boolean functions with this property.

49. When I say that no two gates were identical, I mean that gates at the same position in the two circuits computed different Boolean logic functions. The maximal distance we could achieve in such a random walk depended only little on the frequency of a logic function, that is, on the number of circuits computing this logic function. There may be Boolean logic functions encoded by very few circuits that are highly similar, but those would be very hard to find in a vast circuit space. Previous work has recognized that neutral change can be important for the performance of evolutionary algorithms. See, for example, Banzhaf and Leier (2006), as well as Brameier and Banzhaf (2003), Miller and Thomson (2000), and Yu and Miller (2006). However, to my knowledge, nobody had demonstrated systematically, and in a system that can be implemented in hardware, that genotype networks and the diversity of their neighborhoods are *generic* features of a configuration space (not restricted to one or a few Boolean logic functions).

50. This means that these functions are not likely to be found in the neighborhood of a specific circuit different from the focal circuit in whose neighborhood they occur. (But they may occur in the neighborhood of other circuits in genotype space.)

51. A thousand wiring changes may seem like a lot, but keep in mind that programmable hardware can rewire at lightning speeds. Some useful information on the reconfiguration time of a commercial FPGA is given in Schuck, Haetzer, and Becker (2009).

52. This holds, as usual, for phenotypes (functions) sufficiently frequent that circuits encoding them can be found through a blind search. This property of the circuit library is important, because it can minimize the amount of array reconfiguration (and thus time) needed to change from one function to another.

53. A minor difference from my previous notions of phenotype and function is that any one circuit, any one wiring pattern, may be able to display more than one (computational) behavior, depending on the input to the Boolean logic function that it encodes. One can view a circuit that explores the circuit library to preserve an old, optimal computation while improving a newer, still suboptimal computation as walking along the circuit network in which both computations are unchanged, while exploring the neighborhood of this network for circuits that improve the new computation.

EPILOGUE: PLATO'S CAVE

1. See Darwin (1969), 58.
2. See proposition 168 in Wittgenstein (1983), 99.
3. See Wigner (1960).

4. See Tegmark (2008). A skeptic might argue that the correspondence between mathematics and reality is just an artifact of human history—that there is a huge space of all possible mathematics and we have "sliced" only those theorems from this space that actually describe the physical world. But that assertion raises the question of what the nature of this space is and why useful math exists at all in it.

BIBLIOGRAPHY

Abbott, E. A. *Flatland: A Romance of Many Dimensions*. New York: Perseus, 2002.

Akam, M. "Drosophila Development: Making Stripes Inelegantly." *Nature* 341 (1989): 282–83.

Alberch, P. "From Genes to Phenotype: Dynamical Systems and Evolvability." *Genetica* 84 (1991): 5–11.

Albert, R., and H. G. Othmer. "The Topology of the Regulatory Interactions Predicts the Expression Pattern of the Segment Polarity Genes in *Drosophila melanogaster*." *Journal of Theoretical Biology* 223 (2003): 1–18.

Alfred, R. "Oct. 21, 1879: Edison Gets the Bright Light Right." *Wired*, October 21, 2009.

Amores, A., et al. "Zebrafish Hox Clusters and Vertebrate Genome Evolution." *Science* 282 (1998): 1711–14.

Arkin, R. C. *Behavior-Based Robotics*. Cambridge, MA: MIT Press, 1998.

Arthur, W. B. *The Nature of Technology: What It Is and How It Evolves*. New York: Free Press, 2009.

Ashery-Padan, R., et al. "Pax6 Activity in the Lens Primordium Is Required for Lens Formation and for Correct Placement of a Single Retina in the Eye." *Genes and Development* 14 (2000): 2701–11.

Avery, O. T., C. M. MacLeod, and M. McCarty. "Studies on the Chemical Nature of the Substance Inducing Transformation of Pneumococcal Types: Induction of Transformation by a Desoxyribonucleic Acid Fraction Isolated from Pneumococcus Type III." *Journal of Experimental Medicine* 79 (1944): 137–58.

Babu, M. M., et al. "Structure and Evolution of Transcriptional Regulatory Networks." *Current Opinion in Structural Biology* 14 (2004): 283–91.

Bada, J. L., and A. Lazcano. "Origin of Life—Some Like It Hot, but Not the First Biomolecules." *Science* 296 (2002): 1982–83.

Balch, M. *Complete Digital Design.* New York: McGraw-Hill, 2003.

Banzhaf, W., and A. Leier, A. "Evolution on Neutral Networks in Genetic Programming." In *Genetic Programming Theory and Practice III,* edited by T. Yu, R. Riolo, and B. Worzel, 207–21. New York: Springer, 2006.

Beatty, J. T., et al. "An Obligately Photosynthetic Bacterial Anaerobe from a Deep-Sea Hydrothermal Vent." *Proceedings of the National Academy of Sciences of the United States of America* 102 (2005): 9306–10.

Beerling, D. J., C. P. Osborne, and W. G. Chaloner. "Evolution of Leaf-Form in Land Plants Linked to Atmospheric CO_2 Decline in the Late Palaeozoic Era." *Nature* 410 (2001): 352–54.

Benfey, P., and A. Protopapas. *Genomics.* Upper Saddle River, NJ: Prentice Hall, 2005.

Bennick, A. "Interaction of Plant Polyphenols with Salivary Proteins." *Critical Reviews in Oral Biology and Medicine* 13 (2002): 184–96.

Bergthorsson, U., et al. "Widespread Horizontal Transfer of Mitochondrial Genes in Flowering Plants." *Nature* 424 (2003): 197–201.

Bersaglieri, T., et al. "Genetic Signatures of Strong Recent Positive Selection at the Lactase Gene." *American Journal of Human Genetics* 74 (2004): 1111–20.

Bertalanffy, L. v. *General System Theory: Foundations, Development, Applications.* New York: George Braziller, 1968.

Bharathan, G., et al. "Homologies in Leaf Form Inferred from KNOXI Gene Expression during Development." *Science* 296 (2002): 1858–60.

Blattner, F. R., et al. "The Complete Genome Sequence of *Escherichia coli* K-12." *Science* 277 (1997): 1453–62.

Borges, J. L. *Fictions.* London: Calder, 1962.

Braakman, R., and E. Smith. "The Compositional and Evolutionary Logic of Metabolism." *Physical Biology* 10 (2013): 1–61.

Brakefield, P. M., et al. "Development, Plasticity and Evolution of Butterfly Eyespot Patterns." *Nature* 384 (1996): 236–42.

Brameier, M., and W. Banzhaf. "Neutral Variations Cause Bloat in Linear GP." In *6th European Conference on Genetic Programming (EuroGP 2003),* edited by C. Ryan et al., 286–96. Colchester, England, 2003.

Branden, C., and J. Tooze. *Introduction to Protein Structure.* New York: Garland, 1999.

Brasier, M., et al. "A Fresh Look at the Fossil Evidence for Early Archaean Cellular Life." *Philosophical Transactions of the Royal Society B: Biological Sciences* 361 (2006): 887–902.

Bryson, B. *A Short History of Nearly Everything.* London: Random House, 2003.

Budin, I., R. J. Bruckner, and J. W. Szostak. "Formation of Protocell-Like Vesicles in a Thermal Diffusion Column." *Journal of the American Chemical Society* 131 (2009): 9628–29.

Budin, I., and J. W. Szostak. "Expanding Roles for Diverse Physical Phenomena during the Origin of Life." *Annual Review of Biophysics* 39 (2010): 245–63.

Burchfield, J. *Lord Kelvin and the Age of the Earth*. Chicago: University of Chicago Press, 1990.

———. "Darwin and the Problem of Geological Time. *Isis* 65 (1974): 300–21.

Burr, A. B., and G. E. Andrew. *The Unreasonable Effectiveness of Number Theory*. Washington, DC: American Mathematical Society, 1992.

Bushman, F. *Lateral DNA Transfer: Mechanisms and Consequences*. Cold Spring Harbor, NY: Cold Spring Harbor Laboratory Press, 2002.

Cairns, J., J. Overbaugh, and S. Miller. "The Origin of Mutants." *Nature* 335 (1988): 142–45.

Carroll, S. B., J. K. Grenier, and S. D. Weatherbee. *From DNA to Diversity: Molecular Genetics and the Evolution of Animal Design*. Malden, MA: Blackwell, 2001.

Caspi, R., et al. "The MetaCyc Database of Metabolic Pathways and Enzymes and the BioCyc Collection of Pathway/Genome Databases." *Nucleic Acids Research* 40 (2012): D742–D753.

Cech, T. R. "Structural Biology—The Ribosome Is a Ribozyme." *Science* 289 (2000): 878–79.

Chang, A., et al. "BRENDA, AMENDA and FRENDA the Enzyme Information System: New Content and Tools in 2009." *Nucleic Acids Research* 37 (2009): D588–D592.

Cheng, C. C.-H. "Evolution of the Diverse Antifreeze Proteins." *Current Opinion in Genetics and Development* 8 (1998): 715–20.

Cheng, L. K. L., and P. J. Unrau. "Closing the Circle: Replicating RNA with RNA." *Cold Spring Harbor Perspectives in Biology* 2 (2010).

Ciliberti, S., O. C. Martin, and A. Wagner. "Circuit Topology and the Evolution of Robustness in Complex Regulatory Gene Networks." *PLoS Computational Biology* 3, no. 2 (2007a): e15.

———. "Innovation and Robustness in Complex Regulatory Gene Networks." *Proceedings of the National Academy of Sciences of the United States of America* 104 (2007b): 13591–96.

Cody, G. D., et al. "Primordial Carbonylated Iron-Sulfur Compounds and the Synthesis of Pyruvate." *Science* 289 (2000): 1337–40.

Cohn, M. J., and C. Tickle. "Developmental Basis of Limblessness and Axial Patterning in Snakes." *Nature* 399 (1999): 474–79.

Cole, S., and T. J. Phelan. "The Scientific Productivity of Nations." *Minerva* 37 (1999): 1–23.

Cook, L. M., et al. "Selective Bird Predation on the Peppered Moth: The Last Experiment of Michael Majerus." *Biology Letters* 8 (2012): 609–12.

Copley, S. D. "Evolution of a Metabolic Pathway for Degradation of a Toxic Xenobiotic: The Patchwork Approach." *Trends in Biochemical Sciences* 25 (2000): 261–65.

Copley, S. D., et al. "The Whole Genome Sequence of *Sphingobium chlorophenoli-cum L-1*: Insights into the Evolution of the Pentachlorophenol Degradation Path-way." *Genome Biology and Evolution* 4 (2012): 184–98.

Copley, S. D., E. Smith, and H. J. Morowitz. "The Origin of the RNA World: Co-evolution of Genes and Metabolism." *Bioorganic Chemistry* 35 (2007): 430–43.

Corcos, A., and F. Monaghan. "Role of de Vries in the Recovery of Mendel's Work. 1. Was de Vries Really an Independent Discoverer of Mendel?" *Journal of Heredity* 76 (1985): 187–90.

Corliss, J. B., et al. "Submarine Thermal Springs on the Galápagos Rift." *Science* 203 (1979): 1073–83.

Cotterell, J., and J. Sharpe. "An Atlas of Gene Regulatory Networks Reveals Multiple Three-Gene Mechanisms for Interpreting Morphogen Gradients." *Molecular Systems Biology* 6 (2010): 425.

Cropper, W. H. *Great Physicists*. New York: Oxford University Press, 2001.

Dantas, G., et al. "Bacteria Subsisting on Antibiotics." *Science* 320 (2008): 100–103.

Darwin, C. *On the Origin of Species by Means of Natural Selection, or the Preservation of Favored Races in the Struggle for Life*. 1st ed. London: John Murray, 1859.

———. *The Origin of Species by Means of Natural Selection, or the Preservation of Favored Races in the Struggle for Life*. 6th ed. London: John Murray, 1872. Reprint, New York: A. L. Burt.

———. *The Autobiography of Charles Darwin, 1809–1882*. New York: Norton, 1969.

Davidson, A. R., and R. T. Sauer. "Folded Proteins Occur Frequently in Libraries of Random Amino Acid Sequences." *Proceedings of the National Academy of Sciences of the United States of America* 91 (1994): 2146–50.

Davis, M. C., R. D. Dahn, and N. H. Shubin. "An Autopodial-Like Pattern of Hox Expression in the Fins of a Basal Actinopterygian Fish." *Nature* 447 (2007): 473–76.

Dawkins, R. *Climbing Mount Improbable*. New York: Norton, 1997.

———. *Unweaving the Rainbow: Science, Delusion, and the Appetite for Wonder*. Boston: Houghton Mifflin, 1998.

de Vries, H. *Species and Varieties, Their Origin by Mutation*. Chicago: Open Court Publishing Company, 1905.

Deamer, D. W. "Membrane Compartments in Prebiotic Evolution." In *The Molecular Origins of Life*, edited by E. Brock, 189–205. Cambridge: Cambridge University Press, 1998.

Dekel, E., and U. Alon. "Optimality and Evolutionary Tuning of the Expression Level of a Protein." *Nature* 436 (2005): 588–92.

Delsemme, A. H. "Cosmic Origin of the Biosphere." In *The Molecular Origins of Life*, edited by E. Brock, 100–118. Cambridge: Cambridge University Press, 1998.

Desmond, A., and J. Moore. *Darwin: The Life of a Tormented Evolutionist*. New York: Norton, 1994.

Dong, J., and J. Vagners. "Parallel Evolutionary Algorithms for UAV Planning." In *AIAA First Intelligent Systems Technical Conference*. Chicago, 2004.

Dong, P. D. S., J. S. Dicks, and G. Panganiban. "Distal-less and Homothorax Regulate Multiple Targets to Pattern the Drosophila Antenna." *Development* 129 (2002): 1967–74.

Drake, J. W., et al. "Rates of Spontaneous Mutation." *Genetics* 148 (1998): 1667–86.

Draznin, B. "Molecular Mechanisms of Insulin Resistance: Serine Phosphorylation of Insulin Receptor Substrate-1 and Increased Expression of p85 Alpha—The Two Sides of a Coin." *Diabetes* 55 (2006): 2392–97.

Duarte, N. C., et al. "Global Reconstruction of the Human Metabolic Network Based on Genomic and Bibliomic Data." *Proceedings of the National Academy of Sciences of the United States of America* 104 (2007): 1777–82.

Dubnau, D. "DNA Uptake in Bacteria." *Annual Review of Microbiology* 53 (1999): 217–44.

Eco, U. *Zeichen*. Frankfurt: Suhrkamp, 1977.

Ederer, M. M., et al. "PCP Degradation Is Mediated by Closely Related Strains of the Genus *Sphingomonas*." *Molecular Ecology* 6 (1997): 39–49.

Edwards, J. S., R. U. Ibarra, and B. O. Palsson. "In Silico Predictions of *Escherichia coli* Metabolic Capabilities Are Consistent with Experimental Data." *Nature Biotechnology* 19 (2001): 125–30.

Edwards, J. S., and B. O. Palsson. "The *Escherichia coli* MG1655 in Silico Metabolic Genotype: Its Definition, Characteristics, and Capabilities." *Proceedings of the National Academy of Sciences of the United States of America* 97 (2000): 5528–33.

Eigen, M. "Self-organization of Matter and Evolution of Biological Macromolecules." *Naturwissenschaften* 58 (1971): 465.

Eigen, M., and P. Schuster. *The Hypercycle: A Principle of Natural Self-Organization*. Berlin: Springer, 1979.

Einstein, A. "On the Method of Theoretical Physics." *Philosophy of Science* 1 (1934): 163–69.

Ekland, E. H., J. W. Szostak, and D. P. Bartel. "Structurally Complex and Highly Active Ligases Derived from Random RNA Sequences." *Science* 269 (1995): 364–70.

Eldredge, G., and N. Eldredge. "Editorial." *Evolution: Education and Outreach (Special Issue: Evolution of the Eye)* 1 (2008): 351.

Eng, M. Y., S. E. Luczak, and T. I. Wall. "ALDH2, ADH1B, and ADH1C Genotypes in Asians: A Literature Review." *Alcohol Research and Health* 30 (2007): 22–27.

England, P. C., P. Molnar, and F. M. Richter. "Kelvin, Perry, and the Age of the Earth." *American Scientist* 95 (2007): 342–49.

Espinosa-Soto, C., P. Padilla-Longoria, and E. R. Alvarez-Buylla. "A Gene Regulatory Network Model for Cell-Fate Determination during *Arabidopsis thaliana* Flower Development That Is Robust and Recovers Experimental Gene Expression Profiles." *Plant Cell* 16 (2004): 2923–39.

Feist, A. M., et al. "A Genome-Scale Metabolic Reconstruction for *Escherichia coli* K-12 MG1655 That Accounts for 1260 ORFs and Thermodynamic Information." *Molecular Systems Biology* 3 (2007).

Feist, A. M., et al. "Reconstruction of Biochemical Networks in Microorganisms." *Nature Reviews Microbiology* 7 (2009): 129–43.

Fell, D. *Understanding the Control of Metabolism.* Miami: Portland Press, 1997.

Ferrada, E., and A. Wagner. "Evolutionary Innovation and the Organization of Protein Functions in Sequence Space." *PLoS ONE* 5, no. 11 (2010): e14172.

Ferris, J. P., et al. "Synthesis of Long Prebiotic Oligomers on Mineral Surfaces." *Nature* 381 (1996): 59–61.

Feuda, R., et al. "Metazoan Opsin Evolution Reveals a Simple Route to Animal Vision." *Proceedings of the National Academy of Sciences of the United States of America* 109 (2012): 18868–72.

Finkelstein, A. V. "Implications of the Random Characteristics of Protein Sequences for Their 3-Dimensional Structure." *Current Opinion in Structural Biology* 4 (1994): 422–28.

Fletcher, G. L., C. L. Hew, and P. L. Davies. "Antifreeze Proteins of Teleost Fishes." *Annual Review of Physiology* 63 (2001): 359–90.

Fong, S. S., et al. "Latent Pathway Activation and Increased Pathway Capacity Enable *Escherichia coli* Adaptation to Loss of Key Metabolic Enzymes." *Journal of Biological Chemistry* 281 (2006): 8024–33.

Fong, S. S., and B. O. Palsson. "Metabolic Gene-Deletion Strains of *Escherichia coli* Evolve to Computationally Predicted Growth Phenotypes." *Nature Genetics* 36 (2004): 1056–58.

Fontana, W., and L. W. Buss. "'The Arrival of the Fittest': Toward a Theory of Biological Organization." *Bulletin of Mathematical Biology* 56 (1994): 1–64.

Forster, J., et al. "Genome-Scale Reconstruction of the *Saccharomyces cerevisiae* Metabolic Network." *Genome Research* 13 (2003): 244–53.

Foster, P. L. "Adaptive Mutation: Implications for Evolution." *Bioessays* 22 (2000): 1067–74.

Futuyma, D. J. *Evolutionary Biology.* Sunderland, MA: Sinauer, 1998.

Garcia-Fernandez, J., and P. W. H. Holland. "Archetypal Organization of the Amphioxus Hox Gene-Cluster." *Nature* 370 (1994): 563–66.

Gehring, W. J., and K. Ikeo. "Pax 6—Mastering Eye Morphogenesis and Eye Evolution." *Trends in Genetics* 15 (1999): 371–77.

Gerdes, S. Y., et al. "Experimental Determination and System Level Analysis of Essential Genes in *Escherichia coli* MG1655." *Journal of Bacteriology* 185 (2003): 5673–84.

Gerhart, J., and M. Kirschner. *Cells, Embryos, and Evolution.* Boston: Blackwell, 1998.

Giaever, G., et al. "Functional Profiling of the *Saccharomyces cerevisiae* Genome." *Nature* 418 (2002): 387–91.

Gierer, A., and H. Meinhard. "Theory of Biological Pattern Formation." *Kybernetik* 12 (1972): 30–39.

Gieryn, T. F. *Science and Social Structure: A Festschrift for Robert K. Merton.* New York: New York Academy of Sciences, 1980.

Gilbert, S. F. *Developmental Biology*. 5th ed. Sunderland, MA: Sinauer, 1997.

———. *Developmental Biology*. 9th ed. Sunderland, MA: Sinauer, 2010.

———. "The Morphogenesis of Evolutionary Developmental Biology." *International Journal of Developmental Biology* 47 (2003): 467–77.

Gilbert, W. "Origin of Life—The RNA World." *Nature* 319 (1986): 618.

Goffeau, A., et al. "Life with 6000 Genes." *Science* 274 (1996): 563–67.

Golding, G. B., and A. M. Dean. "The Structural Basis of Molecular Adaptation." *Molecular Biology and Evolution* 15 (1998): 355–69.

Goldschmidt, R. *The Material Basis of Evolution*. New Haven, CT: Yale University Press, 1940.

Goldsmith, T. H. "What Birds See." *Scientific American* 295 (2006): 68–75.

Gomez, C., and O. Pourquie. "Developmental Control of Segment Numbers in Vertebrates." *Journal of Experimental Zoology Part B: Molecular and Developmental Evolution* 312B (2009): 533–44.

Goodman, F. R. "Limb Malformations and the Human HOX Genes." *American Journal of Human Genetics* 112 (2002): 256–65.

Goodman, M., et al. "An Evolutionary Tree for Invertebrate Globin Sequences." *Journal of Molecular Evolution* 27 (1988): 236–49.

Gottesman, M. M., et al. "Genetic Analysis of the Multidrug Transporter." *Annual Review of Genetics* 29 (1995): 607–49.

Gottschlich, D. E., and A. P. Smith. "Convective Heat-Transfer Characteristics of Toothed Leaves." *Oecologia* 53 (1982): 418–20.

Gould, S. J. "Betting on Chance—And No Fair Peeking." In *Eight Little Piggies: Reflections in Natural History*, 396–408. New York: Norton, 1993.

Gould, S. J., and R. C. Lewontin. "The Spandrels of San-Marco and the Panglossian Paradigm: A Critique of the Adaptationist Program." *Proceedings of the Royal Society of London Series B: Biological Sciences* 205 (1979): 581–98.

Gould, S. J., and E. Vrba. "Exaptation—A Missing Term in the Science of Form." *Paleobiology* 8 (1982): 4–15.

Graw, J. "Genetics of Crystallins: Cataract and Beyond." *Experimental Eye Research* 88 (2009): 173–89.

Greenwold, M. J., and R. H. Sawyer. "Linking the Molecular Evolution of Avian Beta (β) Keratins to the Evolution of Feathers." *Journal of Experimental Zoology Part B: Molecular and Developmental Evolution* 316B (2011): 609–16.

Griffiths, A., et al. *An Introduction to Genetic Analysis*. New York: Freeman, 2004.

Guerrier-Takada, C., et al. "The RNA Moiety of Ribonuclease-P Is the Catalytic Subunit of the Enzyme." *Cell* 35 (1983): 849–57.

Haldane, J. B. S. "A Mathematical Theory of Natural and Artificial Selection." *Transactions of the Cambridge Philosophical Society* 23 (1924): 19–40.

———. "The Origin of Life." *Rationalist Annual* 148 (1929): 3–10.

Hall, B. G. "Activation of the bgl Operon by Adaptive Mutation." *Molecular Biology and Evolution* 15 (1998): 1–5.

Hardison, R. C. "A Brief History of Hemoglobins: Plant, Animal, Protist, and Bacteria." *Proceedings of the National Academy of Sciences of the United States of America* 93 (1996): 5675–79.

Harris, G. A. "A New Theory of the Origin of Species." *The Open Court* 18 (1904): 193–202.

Hartl, D., and A. Clark. *Principles of Population Genetics.* Sunderland, MA: Sinauer, 2007.

Hay, A., and M. Tsiantis. "The Genetic Basis for Differences in Leaf Form between *Arabidopsis thaliana* and Its Wild Relative *Cardamine hirsuta.*" *Nature Genetics* 38 (2006): 942–47.

Hayden, E., E. Ferrada, and A. Wagner. "Cryptic Genetic Variation Promotes Rapid Evolutionary Adaptation in an RNA Enzyme." *Nature* 474 (2011): 92–95.

Hedges, S. B., and S. Kumar. "Genomic Clocks and Evolutionary Timescales." *Trends in Genetics* 19 (2003): 200–206.

———. "Precision of Molecular Time Estimates." *Trends in Genetics* 20 (2004): 242–47.

Hersey, G. L., and R. Freedman. *Possible Palladian Villas (Plus a Few Instructively Impossible Ones).* Cambridge, MA: MIT Press, 1992.

Hingorani, M., I. Hanson, and V. van Heyningen. "Aniridia." *European Journal of Human Genetics* 20 (2012): 1011–17.

Hofacker, I., et al. "Fast Folding and Comparison of RNA Secondary Structures." *Monatshefte für Chemie* 125 (1994): 167–88.

Holm, N. G. "Why Are Hydrothermal Systems Proposed as Plausible Environments for the Origin of Life?" *Origins of Life and Evolution of the Biosphere* 22 (1992): 5–14.

Holm, N. G., and E. M. Andersson. "Hydrothermal Systems." In *The Molecular Origins of Life,* edited by E. Brock, 86–99. Cambridge: Cambridge University Press, 1998.

Horowitz, N. H. "The Origin of Life." *Engineering and Science* 20 (1956): 21–25.

Houssaye, A., et al. "Three-Dimensional Pelvis and Limb Anatomy of the Cenomanian Hind-Limbed Snake *Eupodophis descouensi* (Squamata, Ophidia) Revealed by Synchrotron-Radiation Computed Laminography." *Journal of Vertebrate Paleontology* 31 (2011): 2–7.

Hsu-Kim, H., et al. "Formation of Zn- and Fe-Sulfides near Hydrothermal Vents at the Eastern Lau Spreading Center: Implications for Sulfide Bioavailability to Chemoautotrophs." *Geochemical Transactions* 9, no. 6 (2008).

Huang, W. H., and J. P. Ferris. "One-Step, Regioselective Synthesis of Up to 50-mers of RNA Oligomers by Montmorillonite Catalysis." *Journal of the American Chemical Society* 128 (2006): 8914–19.

Hugler, M., et al. "Autotrophic CO_2 Fixation via the Reductive Tricarboxylic Acid Cycle in Different Lineages within the Phylum Aquificae: Evidence for Two Ways of Citrate Cleavage." *Environmental Microbiology* 9 (2007): 81–92.

Huxley, J. *Evolution: The Modern Synthesis.* London: George Allen & Unwin, 1942.

Isalan, M., et al. "Evolvability and Hierarchy in Rewired Bacterial Gene Networks." *Nature* 452 (2008): 840–45.

Jäger, J., et al. "Dynamic Control of Positional Information in the Early Drosophila Embryo." *Nature* 430 (2004): 368–71.

Johannsen, W. L. *Elemente der exakten Erblichkeitslehre.* Jena: Gustav Fischer, 1913.

Johnson, S. *Where Good Ideas Come From: The Natural History of Innovation.* New York: Riverhead, 2010.

Johnston, W. K., et al. "RNA-Catalyzed RNA Polymerization: Accurate and General RNA-Templated Primer Extension." *Science* 292 (2001): 1319–25.

Jörg, T., O. Martin, and A. Wagner. "Neutral Network Sizes of Biological RNA Molecules Can Be Computed and Are Not Atypically Small." *BMC Bioinformatics* 9 (2008): 464.

Jornvall, H., et al. "Zeta-Crystallin versus Other Members of the Alcohol-Dehydrogenase Superfamily—Variability as a Functional Characteristic." *FEBS Letters* 322 (1993): 240–44.

Kacser, H., and R. Beeby. "Evolution of Catalytic Proteins or On the Origin of Enzyme Species by Means of Natural Selection." *Journal of Molecular Evolution* 20 (1984): 38–51.

Kamath, R., et al. "Systematic Functional Analysis of the *Caenorhabditis elegans* Genome Using RNAi." *Nature* 421 (2003): 231–37.

Kapp, O. H., et al. "Alignment of 700 Globin Sequences—Extent of Amino-Acid Substitution and Its Correlation with Variation in Volume." *Protein Science* 4 (1995): 2179–90.

Kappe, G., et al. "Explosive Expansion of Beta Gamma-Crystallin Genes in the Ancestral Vertebrate." *Journal of Molecular Evolution* 71 (2010): 219–30.

Kashefi, K., and D. R. Lovley. "Extending the Upper Temperature Limit for Life." *Science* 301 (2003): 934.

Kauffman, S. A. "Autocatalytic Sets of Proteins." *Journal of Theoretical Biology* 1 (1986): 1–24.

Keats, J. *The Complete Poems of John Keats.* New York: Modern Library, 1994.

Keefe, A. D., and J. W. Szostak. "Functional Proteins from a Random-Sequence Library." *Nature* 410 (2001): 715–18.

Kelley, D. S., et al. "A Serpentinite-Hosted Ecosystem: The Lost City Hydrothermal Field." *Science* 307 (2005): 1428–34.

Kelley, T., and J. Littman. *The Art of Innovation: Lessons in Creativity from IDEO, America's Leading Design Firm.* New York: Crown, 2001.

Kelman, Z., and M. O'Donnell. "DNA Polymerase III Holoenzyme: Structure and Function of a Chromosomal Replicating Machine." *Annual Review of Biochemistry* 64 (1995).

Kenrick, P. "Turning Over a New Leaf." *Nature* 410 (2001): 309–10.

Kern, A. D., and F. A. Kondrashov. "Mechanisms and Convergence of Compensatory Evolution in Mammalian Mitochondrial tRNAs." *Nature Genetics* 36 (2004): 1207–12.

Kettlewell, H. B. D. *The Evolution of Melanism: The Study of a Recurring Necessity.* Oxford: Blackwell, 1973.

Keys, D. N., et al. "Recruitment of a Hedgehog Regulatory Circuit in Butterfly Eyespot Evolution." *Science* 283 (1999): 532–34.

Knight, R. D., S. J. Freeland, and L. F. Landweber. "Rewiring the Keyboard: Evolvability of the Genetic Code." *Nature Reviews Genetics* 2 (2001): 49–58.

Kottler, M. J. "Hugo de Vries and the Rediscovery of Mendel's Laws." *Annals of Science* 36 (1979): 517–38.

Koza, J. R. *Genetic Programming: On the Programming of Computers by Means of Natural Selection.* Cambridge, MA: MIT Press, 1992.

Kreitman, M. "Nucleotide Polymorphism at the Alcohol Dehydrogenase Gene Region of *Drosophila melanogaster.*" *Nature* 304 (1983): 412–17.

Kruger, K., et al. "Self-Splicing RNA: Auto-Excision and Auto-Cyclization of the Ribosomal-RNA Intervening Sequence of Tetrahymena." *Cell* 31 (1982): 147–57.

Kuhn, T. S. *The Structure of Scientific Revolutions.* Chicago: University of Chicago Press, 1962.

Kun, A., M. Santos, and E. Szathmary. "Real Ribozymes Suggest a Relaxed Error Threshold." *Nature Genetics* 37 (2005): 1008–11.

Land, M. F., and R. D. Fernald. "The Evolution of Eyes." *Annual Review of Neuroscience* 15 (1992): 1–29.

Land, M. F., and D.-E. Nilsson. *Animal Eyes.* Oxford: Oxford University Press, 2002.

Lander, E. S., et al. "Initial Sequencing and Analysis of the Human Genome." *Nature* 409 (2001): 860–921.

Lau, K. F., and K. A. Dill. "A Lattice Statistical Mechanics Model of the Conformational and Sequence Spaces of Proteins." *Macromolecules* 22 (1989): 3986–97.

Lawrence, J. G., and H. Ochman. "Molecular Archaeology of the *Escherichia coli* Genome." *Proceedings of the National Academy of Sciences of the United States of America* 95 (1998): 9413–17.

Lawrence, P. A. *The Making of a Fly.* Oxford: Blackwell, 1992.

LeHello, S., et al. "Highly Drug-Resistant *Salmonella enterica* Serotype Kentucky ST198-X1: A Microbiological Study." *The Lancet Infectious Diseases* 13 (2013): 672–79.

Lepland, A., et al. "Questioning the Evidence for Earth's Earliest Life—Akilia Revisited." *Geology* 33 (2005): 77–79.

Levitt, M. "Nature of the Protein Universe." *Proceedings of the National Academy of Sciences of the United States of America* 106 (2009): 11079–84.

Lewin, B. *Genes VI.* New York: Oxford University Press, 1997.

Lewis, E. B. "Gene Complex Controlling Segmentation in Drosophila." *Nature* 276 (1978): 565–70.

Lewontin, R. C., and J. L. Hubby. "A Molecular Approach to the Study of Genic Heterozygosity in Natural Populations, II. Amount of Variation and Degree of Heterozygosity in Natural Populations of *Drosophila pseudoobscura.*" *Genetics* 54 (1966): 595–609.

Liang, Y. H., et al. "The Crystal Structure of Bar-Headed Goose Hemoglobin in Deoxy Form: The Allosteric Mechanism of a Hemoglobin Species with High Oxygen Affinity." *Journal of Molecular Biology* 313 (2001): 123–37.

Lihoreau, M., L. Chittka, and N. L. Raine. "Travel Optimization by Foraging Bumblebees through Readjustments of Traplines after Discovery of New Feeding Locations." *American Naturalist* 176 (2010): 744–57.

Lipman, D., and W. Wilbur. "Modeling Neutral and Selective Evolution of Protein Folding." *Proceedings of the Royal Society of London Series B: Biological Sciences* 245 (1991): 7–11.

Liu, X. Z., et al. 2001. "Avian Haemoglobins and Structural Basis of High Affinity for Oxygen: Structure of Bar-Headed Goose Aquomet Haemoglobin." *Acta Crystallographica Section D: Biological Crystallography* 57 (1991): 775–83.

Lohr, S. "John W. Backus, 82, Fortran Developer, Dies." *New York Times,* March 20, 2007.

Lopes, F. J. P., et al. "Spatial Bistability Generates Hunchback Expression Sharpness in the Drosophila Embryo." *PLoS Computational Biology* 4 (2008).

Loreto, E. L. S., C. M. A. Carareto, and P. Capy. "Revisiting Horizontal Transfer of Transposable Elements in Drosophila." *Heredity* 100 (2008): 545–54.

Lynch, M. *The Origins of Genome Architecture.* Sunderland, MA: Sinauer, 2007.

Lynnerup, N., et al. "Radiocarbon Dating of the Human Eye Lens Crystallines Reveal Proteins without Carbon Turnover throughout Life." *PLoS ONE* 3 (2008): e1529.

Ma, H. W., and A. P. Zeng. "Phylogenetic Comparison of Metabolic Capacities of Organisms at Genome Level." *Molecular Phylogenetics and Evolution* 31 (2004): 204–13.

Martchenko, M., et al. "Transcriptional Rewiring of Fungal Galactose-Metabolism Circuitry. *Current Biology* 17 (2007): 1007–13.

Martin, O. C., and A. Wagner. "Multifunctionality and Robustness Trade-Offs in Model Genetic Circuits." *Biophysical Journal* 94 (2008): 2927–37.

Martin, W., et al. "Hydrothermal Vents and the Origin of Life." *Nature Reviews Microbiology* 6 (2008): 805–14.

Maynard-Smith, J. "Natural Selection and the Concept of a Protein Space." *Nature* 255 (1970): 563–64.

Mayr, E. *The Growth of Biological Thought: Diversity, Evolution, and Inheritance.* Cambridge, MA: Belknap Press of Harvard University Press, 1982.

McCarthy, D. L., A. A. Claude, and S. D. Copley. "In Vivo Levels of Chlorinated Hydroquinones in a Pentachlorophenol-Degrading Bacterium." *Applied and Environmental Microbiology* 63 (1997): 1883–88.

McMahon, J. M., W. L. B. White, and R. T. Sayre. "Cyanogenesis in Cassava (*Manihot esculenta* Crantz)." *Journal of Experimental Botany* 46 (1995): 731–41.

Mendel, G. "Versuche über Pflanzen-Hybriden." *Verhandlungen des Naturforschenden Vereins Brünn* 4 (1866): 3–47.

Merton, R. K. *Social Theory and Social Structure*. New York: Free Press, 1968.

———. "The Unanticipated Consequences of Purposive Social Action." *American Sociological Review* 1 (1936): 894–904.

Miller, J. F., and P. Thomson. "Cartesian Genetic Programming." In *European Conference on Genetic Programming (EuroGP 2000)*, edited by R. Poli et al., 121–32. Edinburgh, Scotland, 2000.

Miller, S. "A Production of Amino Acids under Possible Primitive Earth Conditions." *Science* 117 (1953): 528–29.

Miller, S. L. "The Endogenous Synthesis of Organic Compounds." In *The Molecular Origins of Life*, edited by E. Brock, 59–85. Cambridge: Cambridge University Press, 1998.

Mitchell, M. *An Introduction to Genetic Algorithms*. Cambridge, MA: MIT Press, 1998.

Mjolsness, E., D. H. Sharp, and J. Reinitz. "A Connectionist Model of Development." *Journal of Theoretical Biology* 152 (1991): 429–53.

Moeckel, R., et al. "YaMoR and Bluemove—An Autonomous Modular Robot with Bluetooth Interface for Exploring Adaptive Locomotion." In *8th International Conference on Climbing and Walking Robots (CLAWAR 2005)*, edited by M. O. Tokhi, G. S. Virk, and M. A. Hossain, 685–92. London, 2006.

Mojzsis, S. J., et al. "Evidence for Life on Earth before 3,800 Million Years Ago." *Nature* 384 (1996): 55–59.

Mommsen, T. P., and P. J. Walsh. "Evolution of Urea Synthesis in Vertebrates: The Piscine Connection." *Science* 243 (1989): 72–75.

Monge, C., and F. Leonvelarde. "Physiological Adaptation to High Altitude—Oxygen Transport in Mammals and Birds." *Physiological Reviews* 71 (1991): 1135–72.

Moore, C., and S. Mertens. *The Nature of Computation*. Oxford: Oxford University Press, 2011.

Moran, N. A., J. P. McCutcheon, and A. Nakabachi. "Genomics and Evolution of Heritable Bacterial Symbionts." *Annual Review of Genetics* 42 (2008): 165–90.

Moran, N. A., H. J. McLaughlin, and R. Sorek. "The Dynamics and Time Scale of Ongoing Genomic Erosion in Symbiotic Bacteria." *Science* 323 (2009): 379–82.

Moran, N. A., and A. Mira. "The Process of Genome Shrinkage in the Obligate Symbiont *Buchnera aphidicola*." *Genome Biology* 2 (2001): research0054–research0054.0012.

Morano, I. "Tuning the Human Heart Molecular Motors by Myosin Light Chains." *Journal of Molecular Medicine* 77 (1999): 544–55.

Morgan, T. H. *The Scientific Basis of Evolution*. New York: Norton, 1932.

Morowitz, H. J., et al. "The Origin of Intermediary Metabolism." *Proceedings of the National Academy of Sciences of the United States of America* 97 (2000): 7704–8.

Mrozikiewicz, P. M., et al. "The Significance of C3435T Point Mutation of the MDR1 Gene in Endometrial Cancer." *International Journal of Gynecological Cancer* 17 (2007): 728–31.

Murray, J. D. 1989. *Mathematical Biology.* New York: Springer, 2007.

Nachmann, M. N., and S. L. Crowell. "Estimate of the Mutation Rate per Nucleotide in Humans." *Genetics* (2000): 297–304.

Neidhardt, F. C. *Escherichia coli and Salmonella.* Washington, DC: ASM Press, 1996.

Nelson, K. E., M. Levy, and S. Miller. "Peptide Nucleic Acids Rather Than RNA May Have Been the First Genetic Molecule." *Proceedings of the National Academy of Sciences of the United States of America.* 97 (2000): 3868–71.

Nikaido, M., et al. "Genetically Distinct Coelacanth Population off the Northern Tanzanian Coast." *Proceedings of the National Academy of Sciences of the United States of America* 108 (2011): 18009–13.

Nohynek, L. J., et al. "Description of Four Pentachlorophenol-Degrading Bacterial Strains as *Sphingomonas chlorophenolica* sp nov." *Systematic and Applied Microbiology* 18 (1996): 527–38.

Nüsslein-Volhard, C., and E. Wieschaus. "Mutations Affecting Segment Number and Polarity in Drosophila." *Nature* 287 (1980): 795–801.

O'Brien, P. J., and D. Herschlag. "Catalytic Promiscuity and the Evolution of New Enzymatic Activities." *Chemistry and Biology* 6 (1999): R91–R105.

Ochman, H., and L. B. Jones. "Evolutionary Dynamics of Full Genome Content in *Escherichia coli.*" *EMBO Journal* 19 (2000): 6637–43.

Ochman, H., E. Lerat, and V. Daubin. "Examining Bacterial Species under the Specter of Gene Transfer and Exchange." *Proceedings of the National Academy of Sciences of the United States of America* 102 (2005): 6595–99.

Ochoa-Espinosa, A., et al. "Anterior-Posterior Positional Information in the Absence of a Strong Bicoid Gradient." *Proceedings of the National Academy of Sciences of the United States of America* 106 (2009): 3823–28.

Ogata, H., et al. "KEGG: Kyoto Encyclopedia of Genes and Genomes." *Nucleic Acids Research* 27 (1999): 29–34.

Ogburn, W. F., and D. Thomas. "Are Inventions Inevitable? A Note on Social Evolution." *Political Science Quarterly* 37 (1922): 83–98.

Ohtaka, C., and H. Ishikawa. "Effects of Heat Treatment on the Symbiotic System of an Aphid Mycetocyte." *Symbiosis* 11 (1991): 19–30.

Oparin, A. I. *The Origin of Life.* New York: Dover, 1952.

Oqueta, M., et al. "The Influence of Adh Function on Ethanol Preference and Tolerance in Adult *Drosophila melanogaster.*" *Chemical Senses* 35 (2010): 813–22.

Pal, C., B. Papp, and M. J. Lercher. "Horizontal Gene Transfer Depends on Gene Content of the Host." In *Joint Meeting of the 4th European Conference on Computational Biology/6th Meeting of the Spanish-Bioinformatics-Network,* 222–23. Madrid, 2005.

Panganiban, G., and J. L. R. Rubenstein. "Developmental Functions of the Distal-less/Dlx Homeobox Genes." *Development* 129 (2002): 4371–86.

Pasteur, L. "Des générations spontanées. Soirées scientifiques de la Sorbonne." *Revue des cours scientifiques I* 21 (1864): 258–65.

Piatigorsky, J. "Gene Sharing in Lens and Cornea: Facts and Implications." *Progress in Retinal and Eye Research* 17 (1998): 145–74.

Piatigorsky, J., and G. J. Wistow. "Enzyme Crystallins: Gene Sharing as an Evolutionary Strategy." *Cell* 57 (1989): 197–99.

Poole, A. R., et al. "Composition and Structure of Articular Cartilage—A Template for Tissue Repair." *Clinical Orthopaedics and Related Research* 391 (2001): S26–S33.

Postgate, J. R. *The Outer Reaches of Life*. Cambridge: Cambridge University Press, 1994.

Povolotskaya, I. S., and F. A. Kondrashov. "Sequence Space and the Ongoing Expansion of the Protein Universe." *Nature* 465 (2010): 922–26.

Price, N. D., J. L. Reed, and B. O. Palsson. "Genome-Scale Models of Microbial Cells: Evaluating the Consequences of Constraints." *Nature Reviews Microbiology* 2 (2004): 886–97.

Ptashne, M. *A Genetic Switch: Phage Lambda and Other Organisms*. Cambridge, MA: Cell Press, 1992.

Putman, M., W. H. van Veen, and W. N. Konings. "Molecular Properties of Bacterial Multidrug Transporters." *Microbiology and Molecular Biology Reviews* 64 (2000): 672–93.

Putnam, H. "The Meaning of 'Meaning.'" In *Mind, Language and Reality: Philosophical Papers*, vol. 2: 215–71. Cambridge: Cambridge University Press, 1975.

Radetsky, P. "Life's Crucible." *Earth* 7 (1998): 34–41.

Raman, K., and Wagner, A. "The Evolvability of Programmable Hardware." *Journal of the Royal Society Interface* 8 (2011): 269–81.

Razin, S., D. Yogev, and Y. Naot. "Molecular Biology and Pathogenicity of Mycoplasmas." *Microbiology and Molecular Biology Reviews* 62 (1998): 1094–56.

Rechenberg, I. *Evolutionsstrategie*. Stuttgart: Frommann-Holzboog, 1973.

Redfield, R. J. "Genes for Breakfast—The Have-Your-Cake-and-Eat-It-Too of Bacterial Transformation." *Journal of Heredity* 84 (1993): 400–404.

Rehmann, L., and A. J. Daugulis. "Enhancement of PCB Degradation by *Burkholderia xenovorans* LB400 in Biphasic Systems by Manipulating Culture Conditions." *Biotechnology and Bioengineering* 99 (2008): 521–28.

Reidhaar-Olson, J. F., and R. T. Sauer. "Functionally Acceptable Substitutions in Two Alpha-Helical Regions of Lambda Repressor." *Proteins* 7 (1990): 306–16.

Reidys, C. M. "Random Induced Subgraphs of Generalized n-Cubes." *Advances in Applied Mathematics* 19 (1997): 360–77.

Reidys, C. M., P. Stadler, and P. Schuster. "Generic Properties of Combinatory Maps: Neutral Networks of RNA Secondary Structures." *Bulletin of Mathematical Biology* 59 (1997): 339–97.

Reinhold-Hurek, B., and D. A. Shub. "Self-Splicing Introns in Transfer-RNA Genes of Widely Divergent Bacteria." *Nature* 357 (1992): 173–76.

Reinitz, J., E. Mjolsness, and D. H. Sharp. "Model for Cooperative Control of Positional Information in Drosophila by Bicoid and Maternal Hunchback." *Journal of Experimental Zoology* 271 (1995): 47–56.

Riddle, R. D., et al. "Sonic-Hedgehog Mediates the Polarizing Activity of the ZPA." *Cell* 75 (1993): 1401–16.

Rizzi, M., et al. "Structure of the Sulfide-Reactive Hemoglobin from the Clam *Lucina pectinata*—Crystallographic Analysis at 1.5 Å Resolution." *Journal of Molecular Biology* 244 (1994): 86–99.

Rodrigues, J. F., and A. Wagner. "Evolutionary Plasticity and Innovations in Complex Metabolic Reaction Networks." *PLoS Computational Biology* 5 (2009): e1000613.

———. "Genotype Networks, Innovation, and Robustness in Sulfur Metabolism." *BMC Systems Biology* 5 (2011): 39.

Rosen, W. *The Most Powerful Idea in the World*. Chicago: University of Chicago Press, 2010.

Rothschild, L. J. "The Evolution of Photosynthesis . . . Again?" *Philosophical Transactions of the Royal Society B: Biological Sciences* 363 (2008): 2787–2801.

Russell, P. J. *iGenetics*. San Francisco: Benjamin Cummings, 2002.

Samal, A., et al. "Genotype Networks in Metabolic Reaction Spaces." *BMC Systems Biology* 4 (2010): 30.

Sanchez, R., and J. T. Mahoney. "Modularity, Flexibility, and Knowledge Management in Product and Organization Design." *Strategic Management Journal* 17 (1996): 63–76.

Sanghera, D. K., and P. S. Blackett. "Type 2 Diabetes Genetics: Beyond GWAS." *Journal of Diabetes and Metabolism* 3 (2012): 6948.

Schmitt-Kopplin, P., et al. "High Molecular Diversity of Extraterrestrial Organic Matter in Murchison Meteorite Revealed 40 Years after Its Fall." *Proceedings of the National Academy of Sciences of the United States of America* 107 (2010): 2763–68.

Schopf, J. W., et al. "Laser-Raman Imagery of Earth's Earliest Fossils." *Nature* 416 (2002): 73–76.

Schuck, C., B. Haetzer, and J. Becker. "An Interface for a Decentralized 2D Reconfiguration on Xilinx Virtex-FPGAs for Organic Computing." *International Journal of Reconfigurable Computing* 2009 (2009): Article ID 273791, doi: dx.doi.org/ 10.1155/2009/273791.

Schultes, E., and D. Bartel. "One Sequence, Two Ribozymes: Implications for the Emergence of New Ribozyme Folds." *Science* 289 (2000): 448–52.

Schuster, P. "Prediction of RNA Secondary Structures: From Theory to Models and Real Molecules." *Reports on Progress in Physics* 69 (2006): 1419–77.

Schuster, P., et al. "From Sequences to Shapes and Back—A Case-Study in RNA Secondary Structures. *Proceedings of the Royal Society of London Series B: Biological Sciences* 255 (1994): 279–84.

Schwab, I. R. *Evolution's Witness: How Eyes Evolved*. New York: Oxford University Press, 2012.

Schwartz, J. H. *Sudden Origins: Fossils, Genes, and the Emergence of Species*. New York: Wiley, 1999.

Searcoid, M. O. *Metric Spaces*. London: Springer, 2007.

Sedaghat, A. R., A. Sherman, and M. J. Quon. "A Mathematical Model of Metabolic Insulin Signaling Pathways." *American Journal of Physiology: Endocrinology and Metabolism* 283 (2002): E1084–E1101.

Segre, D., D. Vitkup, and G. Church. "Analysis of Optimality in Natural and Perturbed Metabolic Networks. *Proceedings of the National Academy of Sciences of the United States of America* 99 (2002): 15112–17.

Senecal, P. K., D. T. Montgomery, and R. D. Reitz. "A Methodology for Engine Design Using Multi-dimensional Modelling and Genetic Algorithms with Validation through Experiments." *International Journal of Engine Research* 1 (2000): 229–48.

Sephton, M. A. "Meteoritics—Life's Sweet Beginnings?" *Nature* 414 (2001): 857–58.

Shimeld, S. M., et al. "Urochordate βγ-Crystallin and the Evolutionary Origin of the Vertebrate Eye Lens." *Current Biology* 15 (2005): 1684–89.

Shrestha, B., et al. "Evolution of a Major Drug Metabolizing Enzyme Defect in the Domestic Cat and Other Felidae: Phylogenetic Timing and the Role of Hypercarnivory." *PLoS ONE* (2011): 6.

Sibley, D. A. *The Sibley Guide to Bird Life and Behavior*. New York: Knopf, 2001.

Sim, L., et al. "Structural Basis for Substrate Selectivity in Human Maltase-Glucoamylase and Sucrase-Isomaltase N-Terminal Domains." *Journal of Biological Chemistry* 285 (2010): 17763–70.

Sleep, N. H. "The Hadean-Archaean Environment." *Cold Spring Harbor Perspectives in Biology* 2 (2010): a002527.

Smil, V. *Enriching the Earth: Fritz Haber, Carl Bosch, and the Transformation of World Food Production*. Cambridge, MA: MIT Press, 2000.

Smillie, C. S., et al. "Ecology Drives a Global Network of Gene Exchange Connecting the Human Microbiome." *Nature* 480 (2011): 241–44.

Smith, E., and H. J. Morowitz. "Universality in Intermediary Metabolism." *Proceedings of the National Academy of Sciences of the United States of America* 101 (2004): 13168–73.

Sordino, P., F. Vanderhoeven, and D. Duboule. "Hox Gene-Expression in Teleost Fins and the Origin of Vertebrate Digits." *Nature* 375 (1995): 678–81.

Sproewitz, A., et al. "Learning to Move in Modular Robots Using Central Pattern Generators and Online Optimization." *International Journal of Robotics Research* 27 (2008): 423–43.

Stankiewicz, R. "The Concept of Design Space." In *Technological Innovation as an Evolutionary Process*, edited by J. Ziman, 234–47. Cambridge: Cambridge University Press, 2000.

Stein, U., W. Walther, and V. Wunderlich. "Point Mutations in the MDR1 Promoter of Human Osteosarcomas Are Associated with In-Vitro Responsiveness to Multidrug-Resistance Relevant Drugs." *European Journal of Cancer* 30A (1994): 1541–45.

Steppuhn, A., et al. "Nicotine's Defensive Function in Nature." *PLoS Biology* 2 (2004): 1074–80.

Stevens, M. "The Role of Eyespots as Anti-predator Mechanisms, Principally Demonstrated in the Lepidoptera." *Biological Reviews* 80 (2005): 573–88.

Stevens, M., C. J. Hardman, and C. L. Stubbins. "Conspicuousness, Not Eye Mimicry, Makes 'Eyespots' Effective Antipredator Signals." *Behavioral Ecology* 19 (2008): 525–31.

Stevens, M., C. L. Stubbins, and C. J. Hardman. "The Anti-predator Function of 'Eyespots' on Camouflaged and Conspicuous Prey." *Behavioral Ecology and Sociobiology* 62 (2008): 1787–93.

Stewart, I. *Flatterland.* New York: Perseus, 2001.

———. *In Pursuit of the Unknown: 17 Equations That Changed the World.* New York: Basic Books, 2012.

Stone, J., and G. Wray. "Rapid Evolution of Cis-Regulatory Sequences via Local Point Mutations." *Molecular Biology and Evolution* 18 (2001): 1764–70.

Strandh, S. "Christopher Polhem and His Mechanical Alphabet." *Techniques and Culture* 10 (1987).

Stryer, L. *Biochemistry.* New York: Freeman, 1995.

Sumida, S. S., and C. A. Brochu. "Phylogenetic Context for the Origin of Feathers." *American Zoologist* 40 (2000): 486–503.

Swallow, D. M. "Genetics of Lactase Persistence and Lactose Intolerance." *Annual Review of Genetics* 37 (2003): 197–219.

Szegezdi, E., et al. "Mediators of Endoplasmic Reticulum Stress-Induced Apoptosis." *EMBO Reports* 7 (2006): 880–85.

Szostak, J. W. "The Eightfold Path to Non-enzymatic RNA Replication." *Journal of Systems Chemistry* 3 (2012): 2.

Takiguchi, M., et al. "Evolutionary Aspects of Urea Cycle Enzyme Genes." *Bioessays* 10 (1989): 163–66.

Tamas, I., et al. "50 Million Years of Genomic Stasis in Endosymbiotic Bacteria." *Science* 296 (2002): 2376–79.

Tanaka, F., R. Fuller, and C. F. Barbas. "Development of Small Designer Aldolase Enzymes: Catalytic Activity, Folding, and Substrate Specificity." *Biochemistry* 44 (2005): 7583–92.

Tanay, A., A. Regev, and R. Shamir. "Conservation and Evolvability in Regulatory Networks: The Evolution of Ribosomal Regulation in Yeast." *Proceedings of the National Academy of Sciences of the United States of America* 102 (2005): 7203–8.

Tanenbaum, A. S. *Computer Networks.* Englewood Cliffs, NJ: Prentice Hall, 1988.

Tanner, M. A., and T. R. Cech. "Activity and Thermostability of the Small Self-Splicing Group I Intron in the Pre-tRNA(IIe) of the Purple Bacterium Azoarcus." *RNA-a Publication of the RNA Society* 2 (1996): 74–83.

Tautz, D. "Redundancies, Development and the Flow of Information." *Bioessays* 14 (1992): 263–66.

Taylor, S. V., et al. "Searching Sequence Space for Protein Catalysts." *Proceedings of the National Academy of Sciences of the United States of America* 98 (2001): 10596–601.

Tegmark, M. 2008. "The Mathematical Universe." *Foundations of Physics* 38 (2001): 101–50.

Theissen, G. 2006. "The Proper Place of Hopeful Monsters in Evolutionary Biology." *Theory in Biosciences* 124 (2001): 349–69.

Thomas, G. H., et al. "A Fragile Metabolic Network Adapted for Cooperation in the Symbiotic Bacterium *Buchnera aphidicola*." *BMC Systems Biology* 3 (2009): 24.

Tirosh, I., et al. "A Yeast Hybrid Provides Insight into the Evolution of Gene Expression Regulation." *Science* 324 (2009): 659–62.

Tishkoff, S. A., et al. "Convergent Adaptation of Human Lactase Persistence in Africa and Europe." *Nature Genetics* 39 (2007): 31–40.

Tomarev, S., and J. Piatigorsky. "Lens Crystallins of Invertebrates—Diversity and Recruitment from Detoxification Enzymes and Novel Proteins." *European Journal of Biochemistry* 235 (1996): 449–65.

Tomasz, A. "Antibiotic Resistance in *Streptococcus pneumoniae*." *Clinical Infectious Diseases* 24 (1997): S85–S88.

True, J. R., and S. B. Carroll. "Gene Co-option in Physiological and Morphological Evolution." *Annual Review of Cell and Developmental Biology* 18 (2002): 53–80.

Tucker, V. A. "The Deep Fovea, Sideways Vision and Spiral Flight Paths in Raptors." *Journal of Experimental Biology* 203 (2000): 3745–54.

Tzoulaki, I., I. M. S. White, and I. M. Hanson. "PAX6 Mutations: Genotype-Phenotype Correlations." *BMC Genetics* 6 (2005).

van der Meer, J. R. "Evolution of Novel Metabolic Pathways for the Degradation of Chloroaromatic Compounds." In *Beijerinck Centennial Symposium on Microbial Physiology and Gene Regulation—Emerging Principles and Applications,* 159–78. The Hague, Netherlands, 1995.

van der Meer, J. R., et al. "Evolution of a Pathway for Chlorobenzene Metabolism Leads to Natural Attenuation in Contaminated Groundwater." *Applied and Environmental Microbiology* 64 (1998): 4185–93.

van Ham, R., et al. "Reductive Genome Evolution in *Buchnera aphidicola*." *Proceedings of the National Academy of Sciences of the United States of America* 100 (2003): 581–86.

Venter, J. C. "A Part of the Human Genome Sequence." *Science* 299 (2003): 1183–84.

Vermeij, G. J. "Historical Contingency and the Purported Uniqueness of Evolutionary Innovations." *Proceedings of the National Academy of Sciences of the United States of America* 103 (2006): 1804–9.

von Dassow, G., et al. "The Segment Polarity Network Is a Robust Development Module." *Nature* 406 (2000): 188–92.

Wachtershauser, G. "Evolution of the First Metabolic Cycles." *Proceedings of the National Academy of Sciences of the United States of America* 87 (1990): 200–204.

———. "Groundworks for an Evolutionary Biochemistry—The Iron Sulfur World." *Progress in Biophysics and Molecular Biology* 58 (1992): 85–201.

Waddington, C. H. "Canalization of Development and the Inheritance of Acquired Characters. *Nature* 150 (1942): 563–65.

———. "The Genetic Assimilation of an Acquired Character." *Evolution* 7 (1953): 118–26.

———. *The Strategy of the Genes.* New York: Macmillan, 1959.

Wagner, A. "Redundant Gene Functions and Natural Selection." *Journal of Evolutionary Biology* 12 (1999): 1–16.

———. "Distributed Robustness versus Redundancy as Causes of Mutational Robustness." *Bioessays* 27 (2005a): 176–88.

———. *Robustness and Evolvability in Living Systems.* Princeton, NJ: Princeton University Press, 2005b.

———. "Robustness and Evolvability: A Paradox Resolved." *Proceedings of the Royal Society B: Biological Sciences* 275 (2008): 91–100.

———. "Evolutionary Constraints Permeate Metabolic Networks." *BMC Evolutionary Biology* 9 (2009a): 231.

———. *Paradoxical Life.* New Haven, CT: Yale University Press, 2009b.

———. *The Origins of Evolutionary Innovations: A Theory of Transformative Change in Living Systems.* Oxford: Oxford University Press, 2011.

Watson, J. D., and F. H. Crick. "A Structure for Deoxyribose Nucleic Acids." *Nature* 171 (1953): 737–38.

Whitehead, A. N. *Process and Reality.* Corrected ed. New York: Free Press, 1978.

Wierenga, R. K. "The TIM-Barrel Fold: A Versatile Framework for Efficient Enzymes." *FEBS Letters* 492 (2001): 193–98.

Wigner, E. P. "The Unreasonable Effectiveness of Mathematics in the Natural Sciences." *Communications on Pure and Applied Mathematics* 13 (1960): 1–14.

Williams, L. B., et al. "Birth of Biomolecules from the Warm Wet Sheets of Clays near Spreading Centers." In *Earliest Life on Earth: Habitats, Environments, and Methods of Detection,* edited by S. D. Golding and M. Glikson, 79–112. Dordrecht: Springer, 2011.

Winzeler, E. A., et al. "Functional Characterization of the *S. cerevisiae* Genome by Gene Deletion and Parallel Analysis." *Science* 285 (1999): 901–6.

Wittgenstein, L. *Remarks on the Foundation of Mathematics.* Revised ed. Cambridge, MA: MIT Press, 1983.

Wittkopp, P. J., B. K. Haerum, and A. G. Clark. "Evolutionary Changes in Cis and Trans Gene Regulation." *Nature* 430 (2004): 85–88.

———. "Regulatory Changes Underlying Expression Differences within and between Drosophila Species." *Nature Genetics* 40 (2008): 346–50.

BIBLIOGRAPHY

Wolfenden, R., and Y. Yuan. "Rates of Spontaneous Cleavage of Glucose, Fructose, Sucrose, and Trehalose in Water, and the Catalytic Proficiencies of Invertase and Trehalase." *Journal of the American Chemical Society* 130 (2008): 7548–49.

Wright, P. A., A. Felskie, and P. M. Anderson. "Induction of Ornithine-Urea Cycle Enzymes and Nitrogen Metabolism and Excretion in Rainbow Trout (*Oncorhynchus mykiss*) during Early Life Stages." *Journal of Experimental Biology* 198 (1995): 127–35.

Ycas, M. "A Note on the Origin of Life." *Proceedings of the National Academy of Sciences of the United States of America* 41 (1955): 714–16.

Yu, T., and J. F. Miller. "Through the Interaction of Neutral and Adaptive Mutations, Evolutionary Search Finds a Way." *Artificial Life* 12 (2006): 525–51.

Yus, E., et al. "Impact of Genome Reduction on Bacterial Metabolism and Its Regulation." *Science* 326 (2009): 1263–68.

Zaher, H. S., and P. J. Unrau. "Selection of an Improved RNA Polymerase Ribozyme with Superior Extension and Fidelity." *RNA* 13 (2007): 1017–26.

Zakany, J., and D. Duboule. "The Role of Hox Genes during Vertebrate Limb Development." *Current Opinion in Genetics and Development* 17 (2007): 359–66.

Zhang, X. V., and S. T. Martin. "Driving Parts of Krebs Cycle in Reverse through Mineral Photochemistry." *Journal of the American Chemical Society* 128 (2006): 16032–33.

Zhao, C., et al. "Charcot-Marie-Tooth Disease Type 2A Caused by Mutation in a Microtubule Motor KIF1B Beta." *Cell* 105 (2001): 587–97.

Zimmer, C. *Evolution: The Triumph of an Idea.* New York: HarperCollins, 2001.

INDEX